Driving Intelligence: The Green Book

Driving Intelligence takes a critical and captivating tour of autonomous driving, a phenomenon at the intersection of data-driven platforms, artificial (general) intelligence and the mind. The journey extends from Europe to key areas such as Japan, China, and the US, recognizing the global impact of AI & autonomous driving on high-tech and automotive sectors.

The significance of 'Driving Intelligence' resonates beyond specialized circles, encompassing a spectrum of perspectives – historical, economic, scientific, and philosophical. The book addresses the pressing question of success probabilities and socioeconomic impacts, not just for specialists but for a wider audience keen on understanding the evolution of AI and mobility in the 21st century.

Avoiding partial insights into this domain, the book provides a comprehensive and multifaceted overview which will appeal to a diverse audience including business leaders and policymakers in the mobility and tech industries, governmental bodies, and the general public globally.

Driving Intelligence: The Green Book

Routes to Autonomy

J. Mark Bishop and Gabriel Seiberth

Routledge
Taylor & Francis Group
NEW YORK AND LONDON

Front cover image: Elvira Dratt/Shutterstock

First edition published 2026
by Routledge
605 Third Avenue, New York, NY 10158

and Routledge
4 Park Square, Milton Park, Abingdon, Oxon OX14 4RN

Routledge is an imprint of Taylor & Francis Group, LLC

ISBN: 978-1-032-91122-9 (hbk)
ISBN: 978-1-032-91135-9 (pbk)
ISBN: 978-1-003-56146-0 (ebk)

DOI: 10.1201/9781003561460

Typeset in Nimbus Roman font
by KnowledgeWorks Global Ltd.

Publisher's note: This book has been prepared from camera-ready copy provided by the authors.

The authors acknowledge Philip Dorn & Lukas Milan for their support in the initial phase of the project. We thank Max Haberstroh, Christian Kohlschein, Christian Müller, and Fotis Papastefanou for their sparring and encouragement over the last three years in enabling us to deliver Driving Intelligence. Thank you for your practical advice and insights during the development of these ideas. We also thank Jürgen Schmidhuber, Brad Templeton, Slawomir Nasuto, Kevin Warwick, Stephen Westland, Melanie Mitchell, Jobst Landgrebe, and Max Bennett for insightful comments and targeted recommendations. In addition, we acknowledge our debt to Alex Davies for his excellent book, 'Driven: The Race to Create the Autonomous Car', and Sean Gerrish for his exemplary book, 'How Smart Machines Think', both of which guided and informed our account of the autonomous driving industrys history. We also acknowledge the graphic design expertise of Peta Howlett at iful for producing many excellent illustrations, Thessoloniki's Cafe Baraka, and the help of various AI systems, including Gemini, ChatGPT, Claude, Perplexity, among others. However, any errors that remain are solely the responsibility of the authors, not our ubiquitous AI and human friends. We extend our heartfelt thanks to our families, especially Katerina Koutsantoni & Ada Bishop, and Monika Seiberth, for their patience and support, allowing us to pursue this project on top of the pressures of our day-to-day lives.

Contents

SECTION III Data and Learning

SECTION IV Statistics, Big Data, and Deep Learning

SECTION V Attention, Diffusion, and End-to-end Learning

SECTION VI Two Routes to Autonomy

List of Figures

Foreword

In 2024, more than $150 billion had been invested in autonomous vehicle (AV) technology—yet not a single car could drive itself anywhere a human driver could without intervention.

As an AI researcher and host of *Machine Learning Street Talk* (MLST),[1] who has spent the best part of two decades working with artificial intelligence and having interviewed many of the world's most renowned AI scientists, I have watched the dream of full autonomy evolve and veer. Strangely—after billions of dollars of investment and grandiose hype—we have seemingly failed to bring the technology into existence, much to the chagrin of the Silicon Valley elites!

> 'Although we use the term **embodied** in an **etiolated** form, as there is no **homeodynamic** link between the AI and its tin can vehicular shell, the real world offers far more possibilities than a virtual one. Without **homeodynamics**, however, they are **ontologically** the same' (pp. 21; emphasis added).

The language, as above, can occasionally be a little challenging—and I hope I won't seem overly critical in pointing this out—but I implore the reader to persevere, as the payoff will be significant. Mark and Gabriel's precision with words exemplifies the metaphorical 'war wounds' with which any soldier might be addled from decades of argumentation, reflection, and struggle, having wrangled with these immensely complex scientific and philosophical issues.

Mark and Gabriel concisely point out a significant architectural gap with current AI. Even purportedly 'embodied' and 'causal' AI, which self-driving companies like Wayve in the UK make such a big deal out of in their press releases, is arguably rather weak if its 'brain' is not truly and actively integrated with its robot body in the deep, interactive and self-regulating way which might indeed be necessary for robust adaptation and understanding. Mark has spent much of his academic career expertly teasing out distinctions between existing AI technology and how intelligence works in the biological (translate: real) world. For Mark and Gabriel, an AI robot that is only *extrinsically physically situated* is hardly different from an AI ensconced entirely in a computer simulation—and thus will lack *fundamental properties of intelligence.*

One big shift we have seen in the past decade is a dramatic turn towards end-to-end inscrutable neural networks that learn from vast amounts of data and have very little *hand-crafted* domain-specific knowledge. The cost of this trade-off is that what they gain in their ability to represent gnarly reality, they lose in abstract generalization and compositionally. However, given the incredible progress in AI, it is hard for any critic not to sound like a Luddite, a pessimistic goalpost-mover—affectionately known as 'Cope' for those of you under 30 and *very online.*[2]

Pamela McCorduck captured an aphorism in this direction rather eloquently, and it became known as the 'AI Effect':

'...every time somebody figured out how to make a computer do something—play good chequers, solve simple but relatively informal problems—there was a chorus of critics to say, "that's not thinking".' [235]

In an interview with Wired magazine, Rodney Brooks added:

'Every time we figure out a piece of it, it stops being magical; we say, "Oh, that's just a computation".' [170]

Vehicles which can operate with true autonomy might just be the ultimate test of artificial intelligence—the worthy successor of the Turing Test for the 21st century. [350] Most 'AI benchmarks' are extremely 'gameable', even the much-lauded 'Chatbot Arena', which Cohere researchers recently pointed out was spectacularly gamed by Meta, leading to its Llama 4 launch in March 2025. [376] Subsequently, François Chollet created the 'ARC-AGI benchmark' which resembled a classical intelligence test, and was designed to be 'easy for humans, hard for AI' and to *resist memorisation*—but it too ultimately fell victim to just that when OpenAIs private version of o3 (thought to be o3-pro) scored 87.5% in December 2024—practically human-level performance. [62] Not many seriously think that o3 is generally intelligent, though.

Many efforts have been made to define what exactly artificial *intelligence* is, and the book explores this in excellent detail. For Chollet, it is the efficiency of *knowledge acquisition* or *reasoning*. The researcher Pei Wang defined it in this way:

'Intelligence is the ability of an information-processing system to adapt to its environment with insufficient knowledge and resources.' [370]

Pei is widely regarded as having hit on the most canonical definition of intelligence we currently have. However, he laments the oversimplification through abstraction (losing detail) and idealization (distorting truth) inherent in such a definition, arguing that it should be fruitful enough to guide research. Indeed, in our conversation with Pei on MLST, he cut to the heart of this with a statement that has stayed with me:

'If you really know everything, if you don't have the insufficiency of knowledge and resources, you do not need intelligence.'

This is an interesting reflection given how powerful current AI systems are, even without much (if any) intelligence. Autonomous vehicles falter when faced with the truly open and unpredictable world.

The truth—unwittingly for many—is that current AI systems are a tool, much like Photoshop: They are not really agents. At best, they are shallow simulacra of real agents like us. I remember that, as a child, when I played with Photoshop, I only knew how to use the lens flare effect, and I used it gratuitously, so all the images I created with it looked the same. The reason why 'AI slop' looks like slop is the repetitive, formulaic constructions like '*its not just X ... its Y*'.[3] Good language should be locally unpredictable, yet comprehensible.

It was honest and brave for Mark and Gabriel to acknowledge their usage of LLMs in writing this book, but the reality is that even though much of the AI text that we see online is *slop*, not all of it is. AI might not have autonomy, deep understanding, and compositionality, but its most savvy users can fill in the cracks with iteration and sense-checking and improve the prose. It is possible to have the best of both worlds for many practical purposes when a human is actually in the loop. However, it does raise the question of whether it was always just about aesthetics rather than understanding. The dead internet theory,[4] which sounded far-fetched in 2021 when it was proposed by a user called 'IlluminatiPirate' on an obscure internet forum, [5] has become our *umwelt.* But the unsettling truth is that at least some canny users of generative AI are already using the technology with great facility and evading detection entirely. For the most part, though, social media is stranger than ever, and the fog of war, steeped in this newfangled generative AI, has never been thicker—*but fully autonomous driving really seems to be another matter entirely.*

The Green Book: Routes to Autonomy is a fantastic read and not just a Haynes Manual for AVs ...(*see what I did there?*)[6] It meticulously traces the historical, technical, and philosophical routes taken to date. There are no easy answers, but at least Mark and Gabriel elucidate the true distance yet to be travelled towards genuinely intelligent autonomous systems and thoughtfully map the road ahead.

Dr Tim Scarfe, May 2025.

Notes

1. MLST is the leading technical AI podcast on Spotify: https://www.youtube.com/c/MachineLearningStreetTalk. MLST brings the latest in advanced AI research from leading AI experts around the world. Its approach is unrivalled in terms of its scope and rigour—which is to say, MLST does not just feature research on large language models, but also focuses on other ideas relating to AI from cognitive science, computer science, neuroscience, mathematics, philosophy of mind, and linguistics.

2. Urban Dictionary [Cope]: 'A cope *or coping mechanism* is a psychological strategy whereby someone rejects a harsh truth and adopts a less disturbing belief instead.' [354] See also Nirav Bhan on LinkedIn: 'Cope (noun)—a belief that LLMs cannot reason after one of them cracks open research-level maths.' [27]

3. Wikipedia [AI slop]: 'AI slop, often simply slop, is a derogatory term for low-quality media, including writing and images, made using generative artificial intelligence technology—characterized by an inherent lack of effort, logic, or purpose.' [382]. See also, The Guardian: *Spam, junk ... slop? The latest wave of AI behind the 'zombie internet'*. [137]

4. The 'Dead Internet Theory' asserts that most of the content on the internet today is no longer created by humans but by bots, AI, and automated scripts—and that this shift has been largely hidden from the public. [381]

5. Agora Road's Macintosh Cafe.

6. A cheeky meta-commentary: Pointing out a common, lazy AI writing trope, then using it to praise the book, whilst winking at the reader so they know that this is intentional ...

Preface

In *Driving Intelligence*, we undertake a deep dive into two ostensibly unrelated—and potentially *profoundly* consequential—technological developments of the twenty-first century: autonomous vehicles (AVs) and artificial general intelligence (AGI). AVs are designed to navigate complex real-world environments without human intervention. At the same time, AGI represents a form of artificial intelligence (AI) endowed with *human-level* cognitive abilities—a digital, computational mind capable of performing all cognitive tasks as well as—and in some cases better than—humans.[1]

We do this specifically because we argue that *progress towards full vehicular autonomy* stands as a novel and precise benchmark of *progress towards AGI*. Ultimately, if these technologies can deliver on the hype, the impact will be profound, radically disrupting the way we live and blurring the boundary between humans and machines.

In dissecting such fantasies, the *Driving Intelligence* trilogy, written by established thought leaders in AI and AVs, will unravel the science fiction to reveal the real business case for the big-tech players who—by shifting the capital/labour relationship as mega digital platforms replace markets as the locus of private wealth extraction—increasingly *drive* society.

Artificial general intelligence has historically been an overlooked area of AI research. Although early work in AI aimed at developing systems capable of general, human-level problem-solving, the focus soon shifted towards a narrower, task-based AI, where immediate applications proved more successful. However, recent developments, particularly the rapid advancements made by OpenAI, have reignited interest in AGI and brought the topic from an obscure niche into the limelight of global public interest.

From its formation, OpenAI has consistently aimed to realize AGI by deploying large-scale generative AI via its generative pretrained transformer (GPT) series of chatbots, where it has achieved notable breakthroughs in the field. Thus, today, with companies like Tesla pivoting to adopt a generative AI architecture at the core of its autonomous vehicles,[2] we find the fields of generative AI, computer vision, and autonomous driving converging in the pursuit of Level–5 vehicular autonomy: autonomous driving in all conditions that a human might drive. [104]

As Herbert Roitblat observes, one of the most significant advancements in artificial intelligence in recent years has been the emergence of self-driving cars. [282] Roitblat highlights that real-world problems cannot be easily reduced to formal problems due to the dynamic nature of the world, the imperfection of sensory data, and the lack of clear definitions for system success. He further emphasizes that these challenges make self-driving cars one of the most intriguing computational contexts, standing out from the historical focus of AI research on well-defined, step-by-step problems and solutions.

Adding to this perspective, statistician Gerd Gigerenzer warns that large data sets are only reliable when the future mirrors the past. He cautions that, in the face of surprises, big data—which always draws from historical patterns—may mislead rather than guide future decisions. [116]

Autonomous driving presents artificial intelligence with an exceptionally complex and dynamic environment unmatched by any other use case. Over the past two decades, investments exceeding $150 billion have been directed towards this technology,[3] representing an unprecedented commitment of resources in the scientific realm.[4]

Autonomous driving thus stands as one of the most heavily funded applications of AI in history, serving as a unique test of AI in the open world. Unlike virtual *cognitive* AI applications—such as chatbots that interact in an abstract, controlled domain—autonomous vehicles constitute an *embodied* form of AI,[5] requiring *real-time* interaction with their *real-world* environment.[6]

Proponents argue that while classical AI and algorithmic AI agents are rooted in virtual domains, embodied AI has the potential to become its most valuable and transformative application.[7] By integrating AI into physical systems, autonomous vehicles exemplify this shift, unlocking possibilities far beyond the capabilities of purely cognitive AI.

Tesla set itself the goal of developing a fully self-driving car early on and later supplemented this with the prototype of a humanoid all-purpose robot. The company's CEO Elon Musk has stated that the prerequisite for this is the development of an AI capable of *generalised autonomy*, his paraphrase for *human-level AI*.[8] And we agree that Level–5, the highest level of driving autonomy, certainly does require human-level intelligence as, in simple terms, it entails the ability to operate autonomously at any time and place where a human can drive a vehicle. Thus, we assert that progress towards the development of Level–5 autonomy can be read as a precise benchmark for the development of human-level, generalised artificial intelligence.

Benchmarks have played a pivotal role in AI research from its inception. They are critical research tools that focus attention on the most challenging, unsolved problems.[9] And more: Since intelligence is challenging, if not impossible, to define,[10] the question of when artificial intelligence can be considered achieved has been central to the field since its foundation. Indeed, as early as 1950, Alan Turing proposed a procedure in his paper *Computing Machinery and Intelligence*—later commonly known as the *Turing Test*—which could be used to benchmark machine intelligence qua *thinking*, simply by observing behaviour; an elegant process which neatly sidestepped the need for either a concrete definition of thinking or intelligence.[11]

Since Turing, increasingly sophisticated benchmarks have stimulated progress in AI. They are now almost as important in guiding its development as engineering architectures, curating data, or tweaking algorithms. It is no coincidence that one of the key breakthroughs in computer vision that ultimately enabled autonomous driving—the application of *deep learning* architectures—was prompted by the success of the 'ImageNet Benchmark'.[12]

Recently, the ARC-AGI benchmark (abstraction and reasoning corpus), a set of tasks devised by François Chollet to assess progress towards AGI, has also garnered significant public attention. When OpenAI's o3 system achieved a breakthrough score, some observers were quick to proclaim the arrival of AGI. However, they overlooked the fact that o3 had been pretrained on the task and, in violation of the test rules, relied on extensive computational power during inference to explore a vast latent program space instead of employing abstraction and reasoning to tackle novel challenges.[13] Furthermore, we note that the most recent 'Super Arc' test from Hernandez et al. (2025), based on model abstraction and optimal Bayesian inference for planning, shows '*no clear evidence of LLM convergence towards a defined level of intelligence, particularly AGI or ASI.*' [138]

Thus, while o3's achievement at Chollet's ARC-AGI benchmark was a triumph of sophisticated engineering, it fell short of representing a breakthrough in human-like flexibility in problem-solving, which would require leveraging pre-existing world knowledge—a capability essential for achieving Level–5 autonomy.

Indeed, O3's performance on the ARC-AGI suite has revived the question of whether benchmarks genuinely measure fundamental aspects of intelligence—an assumption that has been repeatedly disproven historically. Given the rapid pace of progress, it becomes increasingly critical to establish benchmarks that meaningfully capture the distinctions between natural and artificial intelligence. However, for one putatively significant difference, *consciousness*,[14] all purely behavioural tests—such as that proposed by Turner and Schneider [351]—remain susceptible to manipulation [31].

In the search for more challenging tests, the focus is shifting from puzzles to real-world problems, such as translating ancient scrolls and developing vaccines, to avoid solutions based on rote memorisation and energy-inefficient brute force computation. Accordingly, a recent Nature article advocates a shift from question-and-answer testing to solving complex tasks under real-world conditions with specific energy efficiency targets. [167] The next frontier for AI testing, according to the article, is the development of benchmarks that assess the ability of AI systems to act as agents in messy and noisy real-world scenarios. These require reasoning under uncertainty and the handling of general queries with multiple complex solution steps and no single correct answer. And this, in turn, requires human-like understanding.[15]

Viewing driving as a proxy for fundamental aspects of general intelligence, such as continual learning, common sense reasoning, intuitive physics, causal reasoning, and theory of mind, we propose that the ability to drive in any human-drivable context serves as a novel benchmark for AGI akin to the Turing Test. While Turing proposed an open—and text-based[16]—conversation as a test for machines' *thinking*,[17] we advocate for *autonomous agency* as a test for a form of AGI that surpasses pattern matching and token memorisation, as it necessitates grounding in the open, real world. Furthermore, the battery-powered nature of modern AVs and their concomitantly limited driving range pose a critical additional *energy efficiency* challenge.

Given that the *real world* is both regular and infinitely variable, there will always be complex new traffic situations—*edge cases*—to navigate. Moreover, since there is only a countably infinite number of Turing machine programs, it is essential that the vehicle's AI can reliably *generalise* across these novel contexts. Indeed, as the aphorism for intelligence attributed to Jean Piaget pithily foregrounds: 'Intelligence is what you use when you don't know what to do'.[18] However, without appropriate contextual *understanding*, such generalisation may not be possible...

Certainly, no current AI system would pass such a *driving test* in an open-ended, real-world scenario of arbitrary complexity. We are missing something important, and our proposed *driving test for AGI* will help narrow down what that might be.[19] Furthermore, as deep learning pioneer Yann LeCun has also hinted that the idea of using autonomous driving as a benchmark for human-level AI may itself help focus research:

> 'How is it that humans can learn to drive a car in about 20 hours of practice with very little supervision, while fully autonomous driving still eludes our best AI systems trained with thousands of hours of data from human drivers? The short answer is that humans rely on their previously acquired background knowledge of how the world works.' [204]

Thus, *Driving Intelligence* will review progress in artificial intelligence through the lens formed by the recent mass-market availability of so-called self-driving cars. Our analysis will shed powerful new light on two popular current narratives that suggest (a) that fully self-driving cars are 'just around the corner' and (b) that AI poses 'an imminent and existential risk to humanity'.

Although such an AGI trenchantly remains some way from being realized, the astounding pace of recent algorithmic developments, coupled with an overarching fear of mankind being subjugated to a mechanical will, prompted the Center for AI Safety (CAIS) to launch an open letter on May 30th, 2023, signed by many of the highest profile figures in AI, warning that the risk posed by AI is *existential*—asserting it is comparable to that of pandemics and nuclear weapons.[20] Hence, in our discussion and assessment of the problematics associated with autonomous driving, our core thesis is that *Driving Intelligence* will also provide a unique perspective on the broader debates currently unfolding around artificial [general] intelligence.

Currently, ChatGPT, Gemini, Claude, and other large language models (LLMs) are trending, with some touting their near-human-like performance in various fields. As the techniques of LLMs—*tokenise everything and scale over data and compute*—are now transferred to automated vehicles, there is an opportunity to liberate agentic AI from the constrained environment of the desktop and to situate it in the open-ended context of an unbounded world. After all, most definitions of intelligence refer to an agent pursuing goals in an open environment, and this is just what autonomous vehicles *do*.

Many multinational companies, especially AI- and data-driven technology giants worldwide, are heavily investing in AVs.[21] If such machines ever become widely available, they will significantly affect many aspects of modern society—promising disruption to employment in the vast global transport sector and ushering in a new era of freedom for people with mental, physical, or age-related difficulties that currently inhibit their mobility. Furthermore, by shifting the capital/labour relationship, as mega digital platforms replace markets as the locus of private wealth extraction, AVs increasingly *drive* society towards what some have labelled a new techno-feudalism. [357]

In the emerging era of self-driving autonomy, vehicle usage will likely prioritise passenger comfort and journey costs over car ownership and the sheer pleasure of driving. The result would be an access economy that manages mobility via massive, globally scalable digital platforms. As DARPA recently reported '*the sensor and electronic box (E-box) combination currently collects four terabytes of sensor data per hour to support AI, machine learning-based autonomy algorithms, and stack approaches*', [81] defining a mega business case that, as a side product, would put unprecedented volumes of 'personal' data at the disposal of those platform players and their data-driven business models. This is the financial and control fantasy that drives much of the enormous global AV investments.

Although autonomous vehicles have yet to fulfil the promises of their most enthusiastic advocates, they are already revealing significant transformative potential today. This is why Tesla's initiative towards 'generalised autonomy' using transformer architectures, the technology behind LLMs like ChatGPT, is garnering so much attention. Indeed, analysts have already suggested that if the company delivers on its promises, it could be positioned to launch a trillion-dollar-scale robotaxi opportunity as early as 2025. [15]

In *Driving Intelligence*, we discuss the successes, difficulties, and fundamental problems around AVs, structured around a traffic-light theme, in three volumes: *Green*, *Amber*, and *Red*. We open this discussion in *The Green Book: Routes to Autonomy*, reviewing the background and some of the early successes of autonomous vehicle technology. Successes which, in 2016, prompted tech drivers and investors such as Elon Musk to confidently predict Level–5 autonomy [automatic driving] to be 'just around the corner'.

Subsequently, *The Amber Book: Errors in Motion, or How Autonomy Fails*, will analyse publicly available data of notable technology failure cases and contextualise these in light of the latest developments in AI that underpin AVs.

Finally, our concluding volume, *The Red Book: Rethinking Autonomy*, will take a look under the hood at the underlying issues that suggest Level–5 autonomy will be extremely hard, if not impossible, to realize safely across all potential driving scenarios—unless there is a radical *a-computational* breakthrough and we realize AGI.

Why *Driving Intelligence*? Few books cater to the general or business reader while offering a deep technical perspective on the broader vision of autonomous driving and artificial general intelligence without succumbing to tech hype. We see these volumes as filling that gap. So join us on this journey along the macadam of vehicular automation as we explore fundamental questions that transcend the boundaries between humans and machines. In doing so, we challenge the very essence of what it means to be human in an age of boundless intelligence, power, and techno-feudalism.

J. Mark Bishop and Gabriel Seiberth, August 2025.

Notes

1. For the latest formalization of AGI, incorporating a Level–based framework akin to the SAE autonomous driving levels, see Morris et al. (2024) [251]. In this model, consistent with SAE classifications, Level–5 AGI is defined as *fully autonomous AGI* or *AI as an agent.*

2. The end-to-end learning approach Tesla adopted in 2024 has been followed by startups like Wayve, Nuro, Waabi, or Ghost Autonomy since 2017. Tesla is emphasized here merely due to its high public profile. [340]

3. According to Automotive News, McKinsey estimates $160 billion of investment in autonomous vehicles over the last two decades. [141]

4. While generative AI has seen remarkable momentum, with investments of $56 billion between 2023 and 2024, mainly due to the success of ChatGPT, launched in November 2022, the cumulative funding for autonomous vehicles remains more than triple this amount, reflecting two decades of sustained investment. In 2025, AI infrastructure spending is projected to reach $250 billion. [183]

5. Although we use the term *embodied* in an etiolated form, as there is no homeodynamic link between the AI and its *tin can* vehicular shell, the real world offers far more possibilities than a virtual one. Without homeodynamics, however, they are ontologically the same. In simpler terms, if we recorded all the sensor data from a vehicle driving in the real world and replayed it in simulation, the controlling AI would not be able to tell the difference. Cf. 'Zombie mouse in a Chinese room'. [257]

6. Wayve Co-founder and CEO Alex Kendall highlights: 'Today's excitement for AI often centres around cognitive AI, like large language models. These are groundbreaking and impactful but remain confined to abstract realities such as words, data, and knowledge. The physical world, where our most significant interactions occur, is the domain of embodied AI, bringing AI's extraordinary potential to real-world interactions.' [180]

7. Yann LeCun, at the World Economic Forum in Davos, predicted a decade of robotics and a new AI revolution based on architectures grounded in the real world. This will enable AI systems to understand, possess persistent memory, reason, and perform complex planning. Unlike processing text, this new paradigm requires AI to develop a mental model of the world from sensory data. [205] Wayve-CEO Alex Kendall asserts: 'embodied AI is poised to become the most valuable application of AI, transforming our interaction with technology and heralding a new era of innovation. Twenty years from now, the majority of people will chiefly associate AI with physical robots that enhance their lives, from household helpers to self-driving cars.' [180]

8. As William Loots remarked: 'In theory, Tesla's anthropomorphic approach appears logical and seems to be the only path to solving AI general vision and general autonomy. In practice, it is several orders of magnitude more difficult to solve without the use of lidar, HD mapping, and radar as training wheels. Still, Musk and Karpathy are correct in asserting that AI general vision is the only possible solution to true autonomy, without the need for HD mapping and localising to a map, as opposed to dynamic and ever-evolving real-world environments.' [215]

   Tesla is not alone in this vision, but it is perhaps the most aggressive. Ghost Autonomy, for example, a startup founded in 2017, backed by OpenAI in 2024 and shutting down only seven months later, called it *universal AI*: Following universal AI principles, its core visual perception neural network, KineticFlow, is not based on image recognition and does not have to be explicitly trained on obstacles to understand and avoid them. [169]

   Other AV 2.0 companies, like Wayve, talk about an etiolated form of embodied AI, [180] in a faint echo of an argument from Hubert Dreyfus: that to have a human-level intelligence requires a human-like body.

9. This is also highlighted by ARC-AGI inventor François Chollet, who states: 'I am a big believer in the idea that the solution to the problem of intelligence must be co-evolved with the challenge—the benchmark.' The benchmark should be a tool that points researchers in the right direction—that is, asking the right questions. But to ask these questions is, in itself, a complex problem.' [63]

10. John McCarthy points out (a) that there is 'no definition of intelligence that doesn't depend on relating it to human intelligence', [233] and (b) the question of what human intelligence really is remains totally open: 'philosophers have not really come to an agreement in 2500 years.' [229] Effectively, McCarthy is conceding that there has been no progress in this area, even though intelligence is at the centre of so many current endeavours, as we also highlight in Section (2.2).

11. Some commentators, for example, Blay Whitby [374], Huma Shah [318] and Geraint Wiggins, (*personal communication*), have suggested that Turing did not intend the imitation game to be the specification of an operational procedure to be performed by future machine intelligence researchers as a yardstick with which to evaluate their wares, but merely as a thought experiment, a '*philosophical ice-breaker .. attempting to deal with the ill-definition ... of the question ... can machines think?*'. The fact that Turing actually enacted the first version of the imitation game offers, perhaps, partial evidence against this interpretation. [348]

12. It is common to refer to AlexNet's performance on ImageNet in 2012 as a key breakthrough because it was the first approach that fully leveraged GPU acceleration. Conversely, some position the first key breakthrough with DanNet, named after Dan Ciresan, in the IJCNN 2011 computer vision competition that used the recognition of traffic signs as a benchmark. For example, Schmidhuber reports that DanNet, which had similarities with the subsequent AlexNet, performed twice as well as humans on the benchmark, three times better than its closest ANN competitor, and six times better than the best non-neural network-based method. [301]

13. OpenAI's o3 *advanced reasoning system* was pretrained on the data and required several hundred times more computational resources than permitted by the rules. [62]

14. This is a critical difference because, as one of the authors of this text has argued in a widely read article on Frontiers, *computation cannot realize consciousness unless a vicious form of panpsychism is true*. [35] We explore this position further in *The Red Book: Rethinking Autonomy*.

15. The question of whether pretrained large language models can be said to 'understand language—and the physical and social situations language encodes—in any humanlike sense', has been extensively discussed by critics such as Mitchell [246] and Bishop [35].

16. By converting the conversation into a text format, the open-ended problem became a task consisting of a finite number of words. Alva Noë highlights this point: 'Although speech is not formally specifiable, text—in the sense of text-messaging—is. Therefore, text can serve as a computationally tractable proxy for genuine human interaction. By filtering all communication between the players through the keyboard, in the name of making the game harder, Turing actually—and really this is a sleight of hand—sweeps what the philosopher Ned Block has called the problem of inputs and outputs under the rug.' [264]

17. As Daniel Dennett wrote: 'Perhaps [Turing] was inspired by Descartes, who in his 'Discourse on Method', plausibly argued that there was no more demanding test of human mentality than the capacity to hold an intelligent conversation.' [83]

18. The full quote is: 'Intelligence is what you use when you don't know what to do: When neither innate ability nor prior learning has prepared you for a particular situation.' Cited after Calvin 1996. [52] The book lacks a specific source, and the cited French work by Piaget does not contain the mentioned remark. Although the quote is repeated by other sources and gained some popularity, they also fail to provide an exact reference, indicating that the quote is likely an urban myth. However, since it offers a concise and memorable summary of Piaget's ideas, we have also chosen to use it.

19. Taking driving as proxy for human-level intelligence is, of course, a—considerable—extension of the famous 'make me a cup of coffee benchmark' by Apple co-founder Steve Wozniak: 'Without prior knowledge of the house, it locates the kitchen and brews a pot of coffee. By this, I mean it locates the coffee maker, mugs, coffee, and filters. It puts a filter in the basket, adds the appropriate amount of grounds and fills the water compartment. It starts the brew cycle, waits for it to complete and then pours it into a mug. This is a task easily accomplished by nearly anyone and is an ideal measure of a general AI.' [300]

20. The CAIS open letter states: 'AI experts, journalists, policymakers, and the public are increasingly dis-

cussing a broad spectrum of important and urgent risks from AI. Even so, it can be difficult to voice concerns about some of advanced AI's most severe risks. The brief statement below aims to address this obstacle and facilitate discussion. It is also meant to create common knowledge of the growing number of experts and public figures who also take some of advanced AI's most severe risks seriously. 'Mitigating the risk of extinction from AI should be a global priority alongside other societal-scale risks such as pandemics and nuclear war'.

21. Five out of the so-called *Magnificent 7*, the seven most-valued public companies that in January 2025 represent more than 40 per cent of the Nasdaq 100 market capitalisation, pursue AV programs or—in the case of Apple—have done this until recently (Alphabet Inc., Apple, Amazon, Tesla, and Nvidia).

Author Biographies

Prof. J. Mark Bishop has been a leading figure in artificial intelligence since the early 1990s. Until his early retirement in 2020, he was Professor of Cognitive Computing and Director of *The Centre for AI and Analytics* at the University of London. Mark has held prominent roles, including Chair of the AISB, the UK's professional body for AI; he currently serves as Chief Scientific Advisor to Fact360, a company specializing in AI-driven cybersecurity solutions. He is also an International Fellow of the Karel Čapek Centre in Prague. Mark's extensive contributions to AI and automated vehicle research include leading the UK team behind the localization module for SENARIO, a pioneering European project on autonomous wheelchair navigation. His work in AI analytics garnered national attention when his research groups tool, *SpendInsight*, was deployed by the UK National Audit Office to uncover potential NHS procurement savings exceeding £500 million per year—an achievement that was reported in the national press and discussed in the UK Parliament.

Mark has provided expert advice on AI policy at the UN, the EU, and the UK Parliament, and has been widely featured in the media, including *Le Monde* and BBC Radio 4s *Agree to Differ*. His critiques of AI have garnered significant attention, notably his article *Artificial Intelligence Is Stupid and Causal Reasoning Will Not Fix It*, which ranks among the most widely read pieces in Frontiers journals. Mark is an influential voice in AI discourse, appearing on international television, including *Voice of America, The Today Programme, Newsnight, Sky News, C4*, as well as high-profile AI podcasts such as *Machine Learning Street Talk*.

Dr Gabriel Seiberth is a distinguished automotive industry professional with 25 years of experience in management consulting and technology. He currently holds a top management position at a leading global electronics and services company for the automotive and mobility ecosystem, which has developed prototypes of automated citybots that integrate into smart city infrastructures. Over the years, he has collaborated with global car manufacturers and leading technology providers on projects for automated driving systems.

Previously, Gabriel served as Managing Director at a leading global technology and consulting firm, where he authored several influential thought leadership pieces. His studies on digital transformation in the automotive industry have been discussed at the German federal government level, and his widely recognised 'Rebooting Autonomous Driving' report on autonomous vehicles gained significant media coverage, including prime-time television in the US. In 2017, he controversially told *Handelsblatt* that Level 5 autonomous driving was as distant as a crewed mission to Mars—a position that, at the time, challenged prevailing industry optimism.

Section I

Background

We open *The Green Book: Routes to Autonomy*, the first volume of *Driving Intelligence*, a trilogy of books exploring how developments in self-driving cars both benchmark and drive progress in artificial intelligence, with a brief sketch outlining the early development of autonomous robots. Following Elon Musk, we argue that if such machines are ever to reach the stage of Level–5 automation—be able to drive anywhere that a human can drive—then computational AI must achieve generalised autonomy; a technological singularity where the computer reaches or exceeds human-level performance across all domains.[1]

Then, to understand how artificial intelligence is realized, we first drill down into the notion of what constitutes *computation* and how the first pioneers of artificial intelligence began to think about building computational systems that could act *intelligently*. Building on this foundation, we will explore how these efforts have progressively led to increasingly complex and sophisticated automated vehicles and examine current trends in the field.

1 Self-Driving to the Future

1.1 A VISION OF AUTONOMOUS TRANSPORTATION

Perhaps a little surprisingly, autonomous vehicles are not just a phenomenon of the 21st century; Leonardo da Vinci, the famous fifteenth-century polymath, designed a self-propelled cart over five centuries ago in 1478.[2] [51] Driven by a spring, da Vinci's wagon was programmed to follow a path across a theatre stage. In principle, such devices could be engineered to follow any fixed, predetermined, traversable course. This concept was subsequently applied in numerous ways, including the development of the compressed-air-engine torpedo. That said, the vision of an autonomous mass transportation system did not emerge until 1939, when there was not even a freeway system in the USA, as Dan McLaughlin revealed in his discussion of the 110 Arroyo Seco Parkway.[3]

Into a seminal exhibition called *Futurama*, an event that was part of New York's 1939 World's Fair, the American industrial designer Norman Bel Geddes projected a realistic model of life in the 1960s, with a particular focus on future transport systems (see Figure 1.1). Visitors rode for almost a third of a mile in audio-equipped chairs across a nearly 36,000-square-foot scale model of a 1960s imagined world, complete with automated highways. Sections included cities, the countryside, and industrial areas. General Motors sponsored the event, and it is widely regarded as marking the birth of mass transit networks and autonomous transportation systems.

At Futurama, Geddes modelled a radical vision of a system of highways with driverless electric cars guided by induction circuits and equipped with radio controls that ensured they maintained a proper distance from each other. Like horse-drawn and train carriages from previous centuries, these vehicle seats were arranged facing each other, allowing passengers to engage in conversation, pastimes, and games more easily. In his vision, Geddes explicitly aimed at removing humans from the driving process with an argumentation reminiscent of today's carmaker communication: 'These cars of 1960 and the highways on which they drive will have in their devices which will correct the faults of human beings as drivers. They will prevent the driver from committing errors.' [22]

Geddes' vision of driverless traffic navigating public roads finally became realized when the United Nations updated the *International Convention on Road Traffic* in 2021 to allow automated driving systems to take over the dynamic driving task instead of a driver. This enables hands-free driving in certain conditions, allowing those in the driver's seat, for example, to use a car's built-in screens to watch movies and TV programmes. Although not proposed until 2021 (83 years after Futurama), critical underpinning technologies towards this goal were developed much earlier.

The first cruise control system was patented in 1945 by a blind and highly ingenious engineer called Ralph Teetor. Teetor was annoyed by the nauseating stop-and-go driving behaviour of his contemporaries—particularly unpleasant for a blind

DOI: 10.1201/9781003561460-2

Figure 1.1 Futurama: highways and horizons. (Photographer unknown, 1939.)

person—and wanted to force them to maintain a steady speed. This led him to design the *Speedostat* system, which deployed a speed selector on the dashboard. Once the selected speed threshold was reached, Teetor's Speedostat mechanism would control the drive shaft to maintain the speed. In 1958, Chrysler was the first company to offer the Speedostat on one of its luxury cars; only one year later, the feature's popularity convinced the carmaker to make it available in all model ranges. In 1960, the feature

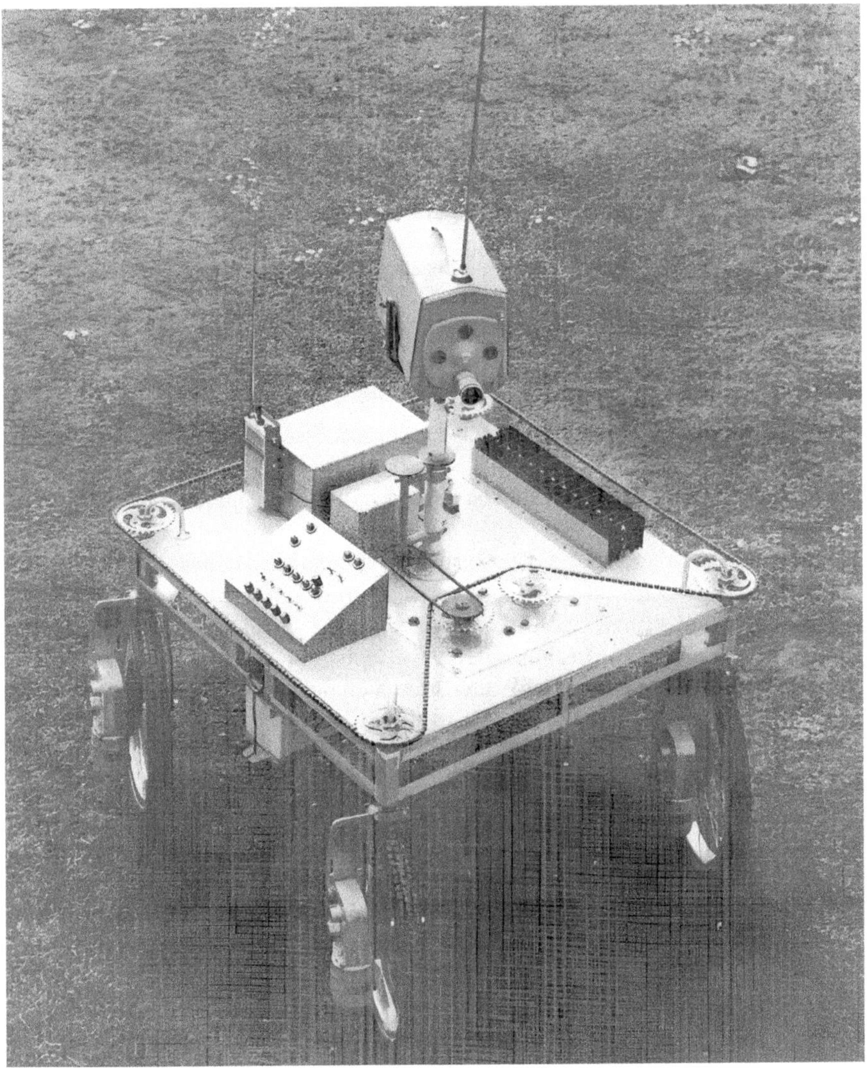

Figure 1.2 James Adams's Stanford cart. (Image adapted from www.CyberneticZoo.com.)

was introduced at General Motors' (GM) Cadillac division and was renamed Cruise Control. The solution was anticipated to save both fuel costs and to help comply with the newly introduced speed limits.

The first use of vision technology in AVs dates from research by Stanford engineering graduate student James Adams in 1961. Prompted by the significant time delay between the moon and earth (which dictated that real-time remote vehicular control was impracticable on the moon), Adams developed the Stanford cart (see Figure 1.2).

The Stanford cart had four small bicycle wheels driven by electric motors and was powered by a car battery. It used a television camera with a fixed view in the forward direction to follow a line drawn on a road. Subsequently, in an article written around 1968, John McCarthy, one of the founding fathers of AI, described this approach to developing computer-controlled cars as requiring: '. . . a computer in the car equipped with a television camera input that uses the same visual input available to the human driver. Essentially, we are proposing an automatic chauffeur.' [230]

1.2 IN THE BEGINNING: WILLIAM GREY WALTER'S *TORTOISE* ROBOTS

In this book, we consider an AV to be a mobile robot that operates without direct human intervention and control. In this sense, the world's first AVs were the robots built in 1948 and 1949 at the Burden Neurological Institute in Bristol, UK, by the left-wing American neuropsychologist and cybernetician William Grey Walter, pioneering research, now viewed as sowing the seeds of future embodied AI and mobile robotics.

Walter's first three-wheeled robots, machines described as *tortoises* because of their shape and slow rate of movement, and a genus he referred to as *Machina Speculatrix*, were named Elmer and Elsie (for **El**ectro**me**chanical **r**obot, light-sensitive). The Burden Neurological Institute (BNI), the institution where William developed his robots, explains the context of Walter's research on their website in the following way:

> 'He [Walter] believed that the secret of how the brain worked lay in how it was wired up and wanted to prove that connections between a small number of brain cells could give rise to very complex behaviours. His first robots, Elmer & Elsie, constructed between 1948 and 1949, were mobile with a plastic shell that was phototropic in that it could follow light and acted as a bump sensor. The robots were designed to show the interaction between two sensory systems: light-sensitive and touch-sensitive control mechanisms (in effect, two nerve cells with visual and tactile inputs). These systems interacted with the motor drive in such a way that the tortoises exhibited behaviour, finding their way around obstacles, for example. They were described as tortoises due to their shape and slow rate of movement and because they *taught us about the secrets of organisation and life*.' [12]

First order cybernetics (FOC) characterises robot-environment systems in terms of feedback loops whose operation can be *interpreted* by an observer in terms of teleological behaviour—i.e. moving towards a goal ['telos']—even though an engineer *designed* the system to *realize* behaviour consistent with her prescribed purpose.

Walter's mobile robots demonstrated that *apparent* teleological behaviours, such as phototaxis—exhibiting behaviours reactive to a visual stimulus, e.g. following a light source without approaching too closely—can be instantiated in a straightforward cybernetic machine (see Figure 1.3). Walter subsequently suggested

Figure 1.3 William Grey Walter's Tortoise robot. (Image adapted from www.CyberneticZo o.com.)

that his robots were, perhaps, the first machines engineered to think and have free will (*sic*) in the manner that biological brains do. In observing their behaviour, Walter famously remarked: 'Despite being crude, the tortoises conveyed the impression of having goals, independence, and spontaneity.'[4]

1.3 FROM SHAKEY TO HERBERT: TOP-DOWN AND BOTTOM-UP PARADIGMS FOR AUTONOMOUS VEHICLE CONTROL

While Walter's tortoise machines can be viewed as pioneering AVs, they were not practical devices in and of themselves, as they could not easily be deployed to perform many valuable tasks (say, to navigate a cargo from point A to point B). In marked contrast, in the 1960s, at the Stanford Research Institute (SRI) in Menlo Park—now SRI International—an alternative approach to building more controllable mobile robots began to be developed.

The Shakey project (see Figure 1.4) took place from 1966 to 1972 at the SRI and rapidly emerged as a seminal contribution to AI and autonomous robotics. [191] Shakey combined research in AI, robotics, computer vision, and natural language processing to analyse commands, breaking them down into component elements to *reason* about potential actions. In this way, Shakey *perceived* the world and *planned* how to *act* to achieve a specified *goal*.

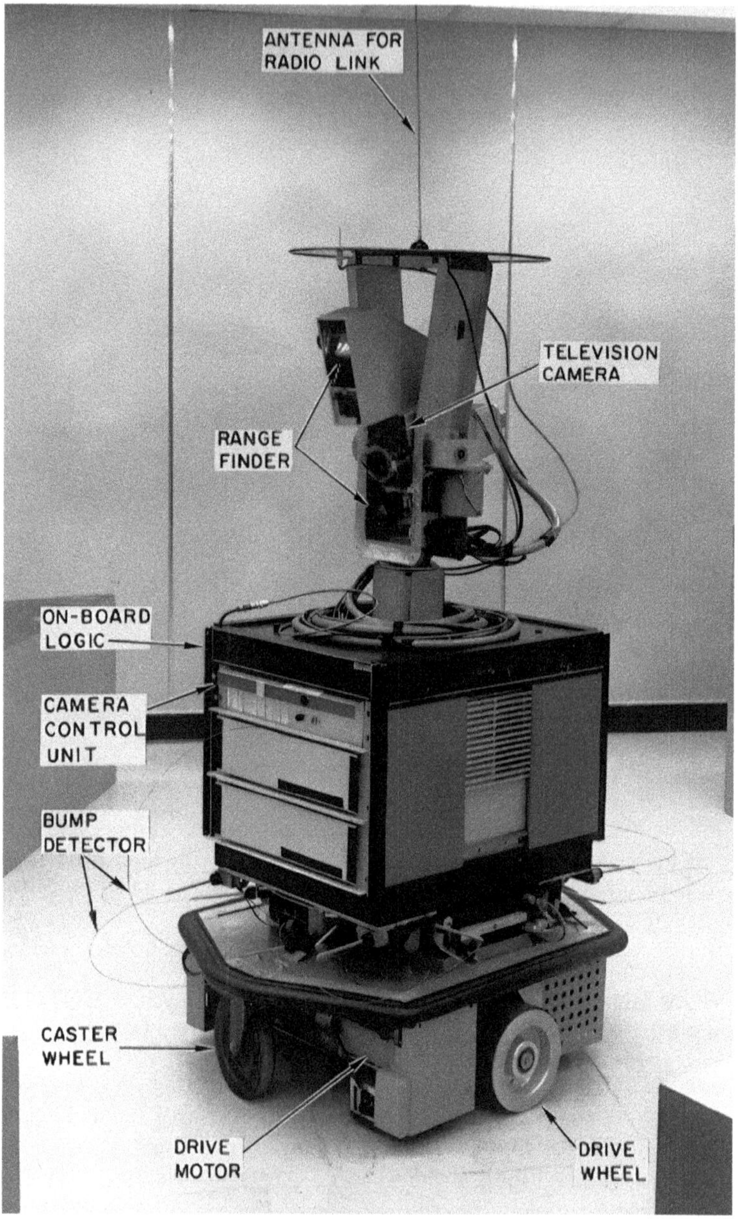

Figure 1.4 Shakey the robot. (Image adapted from Wikipedia: *Shakey the robot* and licensed under the Creative Commons Attribution-Share Alike 3.0 Unported license: https://creativecommons.org/licenses/by-sa/3.0/deed.en.)

In contrast to other contemporary approaches to mobile robotics, where systems had to be painstakingly instructed to execute each of the individual steps involved in completing a task, Shakey's AI and sense, plan, act (SPA) cycle was radically *autonomous*.[5] And at its heart lay the [philosophical] notion of a machine building a rich *internal* representation of the *out-there* physical world, a representation that the machine could use in its subsequent reasoning and acts.

From a historical perspective, Shakey's pioneering implementation of the so-called top-down, or hierarchical, control paradigm is the oldest and best-known method of organizing and deploying intelligence in mainstream autonomous robotics. It dominated mobile robotics research from 1966 until the late 1980s when Rodney Brooks' radical *reactive robotics* paradigm burst onto the scene.

In contrast to Shakey, Brooks' *direct perception* approach was based on the insight that it was inherently much more difficult to sense the world to build the complex internal *representations* that AI could use to reason and act than to 'use the world as its own best representation'; in Brooks' iconoclastic vision, the robot merely needs to sense the world as and when information about it is required. [48]

Both approaches have shaped the design of autonomous vehicles. However, by rejecting reliance on detailed internal representations (high-precision maps or programmed traffic rules) and embracing the idea that the world serves as its own best representation, which AV cameras perceive and process on the fly, Elon Musk's vision to revolutionize the robotaxi industry exemplifies the latter. We will revisit this dichotomy in detail later in our journey.

Viewed in a historical light, there are strong analogies between Brooks' core insight—*to use the world as its own best representation and to sense the world, as required to obtain further information from it*—and the subsequent development of the *active vision* paradigm in computer vision, wherein the viewpoint of the visual sensor(s) is manipulated to sense the environment and solicit better information from it. In this manner, active vision develops a theme initially explored in cognitive science [as animate vision] by Hana Ballard and in computer vision by Aloimonos et al. [7]; as Hana explains:

> 'We are accustomed to thinking of the task of vision as being the construction of a detailed representation of the physical world. Furthermore, this constructive process is regarded as being independent of larger tasks. From the Encyclopedia of Artificial Intelligence, *the goal of an image understanding system is to transform two-dimensional data into a description of the three-dimensional spatiotemporal world* and such a system *must infer 3-D surfaces, volumes, boundaries, shadows, occlusion, depth, colour, motion*. However, a paradigm that we term *animate vision* argues that vision is more readily understood in the context of the visual behaviours that the system is engaged in and that these behaviours may not require elaborate categorical representations of the 3-D world.' [21]

Building massive, high-fidelity representations of the out-there world is a highly complex process. Achieving this without a substantial lag between the robot's position in the world and its derived representation requires significant computational resources. An immediate benefit of Brooks' insight, particularly relevant in earlier decades when requisite computing power was prohibitively expensive, is that by directly sampling the world, such systems sidestep practical issues of *sense lag*, wherein the relationship of the robot to the world changes before its internal model can be updated and re-computed. Hence, robots built using a reactive architecture require relatively modest processing power and are much easier to operate at relatively high-speed compared to those engineered using the top-down representational paradigm.

Indeed, when one of the first robots built using Brooks' architecture by Kevin Warwick and colleagues at the University of Reading, UK, was first powered up, it proceeded to drive across the lab floor so rapidly that it was unable to stop safely. An unplanned altercation with a nearby corridor wall ensued, terminating the vehicle's research lifespan somewhat sooner than had been hoped.

To enable reactive robots built with behaviour-based subsumption architecture to perform complex tasks, Brooks rejected the concept of a centralized coordinator directing actions, such as a controller managing a sense–plan–act (SPA) cycle. Instead, he proposed a population of layers operating independently and in parallel, each responsible for a specific behavioural goal. Lower-level behaviours, such as wandering around, could be overridden or *subsumed* by higher-level behaviours, like navigation, based on the robot's internal state and sensory data. In this way, Brooks' approach eliminated the need for a centralized hierarchy or overarching control, an idea that would prove to be very forward-looking and influential. We will revisit this central question—central control or independent coordination/subsumption—multiple times during our endeavour.

One slightly frivolous example of the reactive robotics paradigm, designed by Jonathan Connell at MIT in 1980, was a minimalist robot called 'Herbert' (see Figure 1.5). Herbert was developed to function according to subsumption principles and, famously, was able to successfully navigate around the lab, collecting researchers' discarded empty soda cans. [70] However, Herbert's impressive can-collecting abilities also exposed the limitations of subsumption architecture. The robot was unreliable, with no record of flawlessly completing its task, and no purely subsumption-based robot has since matched its capabilities. [112]

1.4 THE FIRST AUTONOMOUS CAR

However, it wasn't until a decade after the Shakey robot's first tentative steps in reasoned motion that research towards building self-driving passenger vehicles began in earnest. In 1977, Sadayuki Tsugawa and his colleagues at the University of Tsukuba's Mechanical Engineering Lab in Japan built the world's first autonomous intelligent car (see Figure 1.6). Tsugawa's vehicle, using two video cameras and analogue signal processing to detect street markings, was reported driving autonomously on roads at speeds of up to 30 km/h.

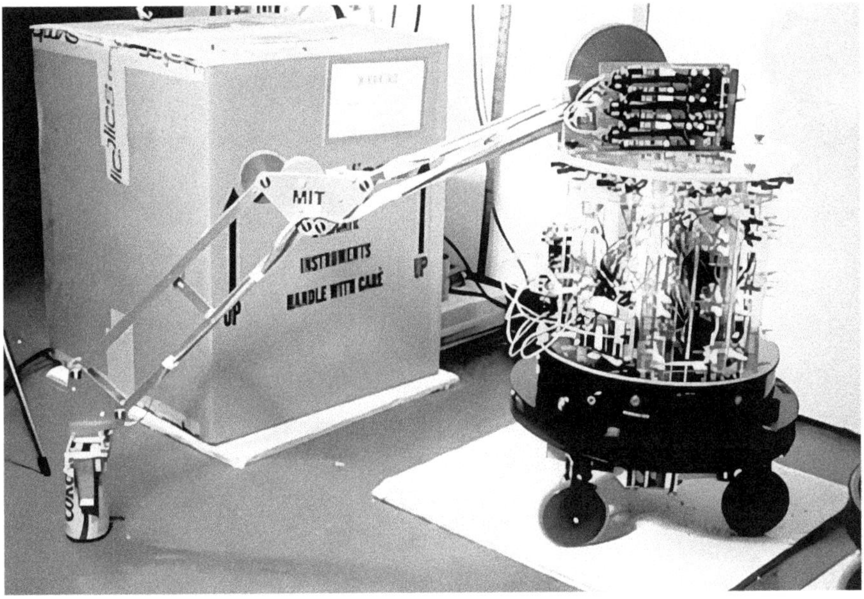

Figure 1.5 Jonathan Connell's Herbert the soda-can-collecting robot. (Image adapted from Britannica: *nouvelle artificial intelligence* [98].)

Tsugawa and colleagues described their invention in a seminal paper from 1979. [347] At first sight, their description sounds very much like the autonomous cars we know today:

> 'The vehicle is completely **autonomous** and can be driven without a human driver. The road pattern recognition unit, involving a pair of TV cameras and a processing unit, identifies obstacles in front of the vehicle and outputs data regarding the locations of the obstacles. The problem-solving unit is a micro-computer system that determines optimal control of the environment around the vehicle based on the data.' [347]

However, although Tsugawa's intelligent car was able to drive on the road without human intervention, its performance fell far short of that of today's AVs. Nevertheless, Tsugawa's intelligent car remains of significant historical interest, not least as an astounding celebration of what can be achieved by software in 8 kB of RAM.[6]

A decade later, Tsugawa's broad approach, which predominantly followed the road by identifying characteristic road markings, was perfected by the German engineer Ernst Dickmanns, who is sometimes dubbed the man who invented the self-driving car. [82] Dickmanns deployed an array of cameras and 60 microprocessors mounted within a specially engineered Mercedes van to detect relevant objects on the road in front and behind the vehicle. It undertook this task after actively filtering

Figure 1.6 Sadayuki Tsugawa's intelligent car. (Image adapted from Dickmanns, 2020 [88].)

out distracting noise in the camera data—i.e. Dickmanns' systems focused on attending to objects of immediate relevance to driving, contra building a rich 3D model of *everything* in the visual scene.[7]

Subsequently, in 1986, *VaMoRs* (Versuchsfahrzeug für autonome Mobilität und Rechnersehen—*Test vehicle for autonomous mobility and computer vision*), pioneered and developed by the team around Dickmanns, became the first vehicle to drive autonomously on a track at Bundeswehr University Munich (Uni-BwM). The following year, VaMoRs was driven down an empty section of a Bavarian autobahn at speeds approaching 60mph. This marked the first in a series of milestones in the development of autonomous transportation. [97]

One such milestone was Dickmanns' demonstration of a highway autopilot in 1994. Dickmanns, funded by the European PROMETHEUS program, developed two autonomous prototypes, VAMP and VITA-2, based on a Mercedes S-class.[8] These advanced research cars, equipped with a highway autopilot, successfully demonstrated their capabilities on a 1994 test run from Munich to Denmark and back, performing lane changes and overtaking at 75 mph (110 km/h). This project was a fascinating precursor to the Level–3 system Mercedes introduced in its current S-Class 20 years later.

Nonetheless, even after the huge step forward established by his pioneering approach, Dickmanns subsequently underscored the need to wait for more powerful computers (better hardware) to drive more intelligent behaviours (better AI algorithms; better software), observing, over the next decade, the field of AD slowly begin to *hibernate in a virtual AI winter* as these waited to be developed. [113]

Dickmanns' thinking was visionary and forward-looking. [89] In the course of his work, he piloted pioneering methods such as saccadic vision, probabilistic reasoning (utilizing Kalman filters, as they are still used in AVs today) and even parallel computing. His approach, which he called *dynamic vision*, anticipated key elements of the Tesla strategy, which—much later—would earn the electric car pioneer a fabulous valuation on the capital markets. In addition, Dickmanns, especially interesting for our context, explicitly described *dynamic vision* as a key element for artificial general intelligence (AGI) as early as 2011. [87]

At the *4th Conference on Artificial General Intelligence* in 2011, [311], the German professor gave a talk that farsightedly proposed fast deep nets for AGI vision with biologically inspired adaptive vision models, [310] preceding Elon Musk's positioning of FSD as *generalized autonomy* by many years. The engineer even anticipated Tesla's argument that camera-based dynamic vision is more efficient than sensing with lidar and radar, and is possible because humans also perceive this way.[9]

Only two years after VaMoRs' successful test drive at Bundeswehr University, graduate student Dean Pomerleau, from Carnegie Mellon University (CMU) in Pittsburgh, experimented with a radically different *machine-learning* approach. This move proved equally prescient, albeit it was just as restricted by hardware capabilities as that of his German predecessor. In his 1988 paper 'ALVINN: An Autonomous Land Vehicle in a Neural Network', [275] Pomerleau described how a three-layer neural network could be trained for the task of road following. It was one of the first applications of the newly operationalized machine learning approach that, with more powerful computation, was to revolutionize AI just a decade or so later.

What was novel about Dean's approach was not just its use of a neural network but also how the neural network was applied. Pomerleau employed machine learning to develop a system that controlled the entire processing pipeline required to steer a vehicle. This approach, now known as *end-to-end* learning, abandoned the prevailing modular sense–plan–act architecture and involved feeding a digitized image directly into a neural network, which immediately generated steering-control output signals.[10] The instruction to the system can be described as something like: 'If the image looks like this, turn the wheel like that.' This is another core idea that we will encounter in variations throughout our book.

To achieve this remarkable feat, a camera system was trained using supervised learning on a dataset of 1,200 simulated road images to detect the centre of the lane on the road. Subsequently, Pomerleau applied online *imitation learning* using data logged from a human driving the vehicle. Once the camera system had identified the road, it could connect to the steering system to control the car. To simplify things, the system was designed to ignore complications like oil slicks, ruts, tyre marks or dynamic weather effects.

The process proved highly effective, though the system occasionally lost self-driving guidance when transitioning between different types of roads. [270]

Nonetheless, the absence of a '3D map' significantly enhanced its adaptability, allowing it to operate successfully in previously unseen environments.

The test vehicle, a camouflaged army ambulance named NAVLAB, initially had a top speed of 3.5 mph, constrained by its limited computing power. Its CPU, the size of a refrigerator, operated at 100 million floating point operations per second and required a 5 kW generator to power it. [132] Nonetheless, following several years of development, the system eventually achieved speeds of up to 70 mph on some (wide) roads.

On July 23, 1995, Pomerleau and CMU PhD student Todd Jochem embarked on a nine-day journey across the United States in NAVLAB-5, an autonomous minivan, as part of the 'no hands across America' project. The goal was to demonstrate a new road-following system called RALPH (Rapidly Adapting Lateral Position Handler). [274] Unlike ALVINN's end-to-end neural network controller, RALPH broke the steering task into three steps: sampling the image, determining road curvature, and calculating the vehicle's lateral offset relative to the lane centre. The outputs of the latter two steps were combined to generate a steering command, which could either serve as a road-departure warning system by comparing it to the human driver's input or be sent directly to NAVLAB-5's steering motor for autonomous control. The two engineers managed the car's accelerator and brake controls while the system autonomously navigated 2,797 miles (98.2% of the trip) across ten states and multiple cities, relying on computing power equivalent to an Apple Watch.[11]

'No hands across America' marked the first long-duration field test of a self-steering car and laid the groundwork for CMU pioneers such as William Red Whittaker and Chris Urmson. Their subsequent work developed these early approaches into the landmark DARPA Challenge vehicles (as detailed in Section 5). Later, as the leader of Google's self-driving car project, Urmson was tasked with pioneering the deployment of AV technology in real-world scenarios.

Although initially successful, ALVINN's end-to-end machine learning system ultimately proved less effective than multi-step modular approaches (such as Pomerleau subsequently developed in RALPH). Nonetheless, interest in such end-to-end control never wholly waned, experiencing a dramatic resurgence just a decade or so later.

In 2003, Yann LeCun and collaborators developed an autonomous off-road vehicle control system using end-to-end machine learning. Funded by the Defense Advanced Research Projects Agency (DARPA), the project became known as DARPA Autonomous Vehicle (DAVE). [203] In 2015, Nvidia introduced DAVE-2 to demonstrate how a system could learn to detect road boundaries without human-labelled data during training. Instead, the system extracted meaningful road features directly from a sparse training signal, using only steering data. [40] The Berkeley *deep drive project* took this approach further in 2017 and extended it towards multi-modal and multi-task learning. Lead developer Karl Zipser and colleagues used 1:10 scale model cars with Nvidia Jetson mini-computers and an active vision system. The neural networks were trained on 150 hours of human-controlled driving data using behaviour cloning. The model cars were able to follow high-level directional signals to navigate a complex path in unknown real-world environments, both on streets and off-road, including steep climbs and downward slopes. [64] [153] Energy consumption is a notable feat of this setup: the Jetson mini-computer requires just 5–20 W, while current autonomous driving chips require 700 to 1,200 W.

Now, fast-forward to 2024, when Tesla launched the new Full Self-Driving (FSD) system based on an end-to-end neural network. Instead of using a modular system with handcrafted rules in the planning module, Tesla now uses its wealth of driving data to train the whole data pipeline by imitating human driving behaviour obtained from millions of videos.

The drives on all these videos were graded, and the network was trained to imitate the actions of the good drivers. The benefits are substantial: the models can generalize better when facing unknown scenarios by finding the closest good behaviour in the dataset to adopt. With this, Tesla began pivoting away from convolutional neural networks based on supervised learning and towards the technologies that power LLMs like ChatGPT: generative AI, transformers, self-attention mechanisms and auto-regression.

As the signals from an AV's sensor can be mapped to strings of tokens similar to those used when ChatGPT processes language, a transformer-based end-to-end controller that has been pretrained on millions of good driving examples can often exhibit excellent, almost human-like driving behaviour.[12] Hence, Tesla is aggressively pushing to achieve full self-driving and generalized autonomy qua AGI via an active vision system that is effectively based on end-to-end neural network control and camera sensors.

1.5 AGI: FROM HERE TO ETERNITY

Just imagine, for a moment, a future where your mind *can* be functionally scanned and uploaded to a computer—at a stroke, jacking you into a brave new digital world, free from the constraints of your frail biological body, to live for eternity in a boundless, psychedelic, silicon heaven. This *is* your future, according to some leading transhumanists and futurologists who are confident that the human race will rapidly evolve beyond its current physical and mental limitations, thanks to an inexorable triumvirate of science, engineering, and computation.

In a bizarre twenty-first-century echo of the ancient pharaohs, who believed that their souls would exist forever as long as their *name* was never forgotten, there is a growing realization that the human-level artificial intelligence required for Level–5 autonomous driving holds within its computer technology the science fiction promise of mind existing for eternity, *just as long as its digital twin is never forgotten.*[13]

And so begins a wild fantasy that takes us far beyond the limits of our current understanding of what it means to be human as we enter a new 'Age of Aquarius', a new digital era of 'peace, love, and understanding'. Such *mind uploading* will be the next great leap forward for humanity—or so this explicitly dualist, quasi-religious, death-denying fantasy runs.

This vision of mind uploading is predicated on the intuition that a suitably programmed digital computer can replicate *all* aspects of the human mind—i.e. the uploaded mind, now instantiated by computer, can perform [at least] as well as the human, across *all* the myriad tasks that biological humans can perform. Thus, at the heart of its conceit, mind uploading entails that artificial general intelligence is both theoretically possible and technically realizable. This is a core belief shared by at least one notable figure who significantly features in our discussion of AVs, Mr Elon Musk.

Elon Musk is an American entrepreneur and businessman, and, during a brief period in 2025, an official of the Trump administration. He is the founder, CEO, and CTO of SpaceX, CEO of Tesla, Inc., founder of X.com—an online payment company with a broad vision of revolutionizing the financial industry, which later became PayPal—and most recently, founder and sole director of xAI. Musk has stated on numerous occasions that he believes that mind uploading is possible and that it is something that society should be working towards. [254] And, it transpires, there is a very good reason why Mr Musk is such an ardent promulgator of mind-uploading and AGI because, as we shall see in this book, AGI is an engineering prerequisite to achieve full Level–5 AD.

Indeed, in 2021, Elon Musk publicly concluded that the only path to fully self-driving cars, unrestricted (Level–5) autonomy, requires the development of an AI system that is—like humans—able to solve all challenges it encounters autonomously, effectively specifying a requirement for Tesla to develop AGI:

> 'While perhaps obvious in hindsight, it turns out that the only way to solve self-driving is to solve a big part of real-world AI. That is what Tesla is doing, both in hardware and software. Unless a company is good at AI with tight compute, it's hopeless. This is insanely hard.' [253]

Thus, in Musk's neo-ecclesiastical vision, the road to AVs is **driving—*artificial*—intelligence** towards AGI, and in the process, quite literally, driving humanity to eternity.

However, as we will uncover in this trilogy of books, there are many deep-rooted reasons why realizing AGI will be extremely hard to achieve. Indeed, these reasons lead us to doubt that AGI will arrive anytime soon, *if it ever does at all*; a conclusion that has global implications for the future of both the AI and AD industries and, surprisingly, also the chimaerae of mind uploading and digital immortality.

Notes

1. OpenAI's CEO, Sam Altman, asserted on January 6, 2025, that AGI is essentially resolved ('We are now confident we know how to build AGI as we have traditionally understood it'), without specifying the definition he is using. His intention in declaring the mission accomplished is evidently to shift the company's resources to the subsequent phase, which he terms *Superintelligence*, again without providing a clear definition. [271]

2. To clarify, **all** so-called AVs are, sensu stricto, **allopoietic** devices, in that they are all systems engineered to fulfil a *human* purpose; sensu stricto none are fully autonomous, capable of their independent teleological behaviour (fortunately, for the likes of Tesla, Waymo, etc.).

3. 'Its construction began on March 22, 1938, under the direction of Spencer V. Cortelyou and A.D. Griffin (district chief engineer and design engineer, respectively). Although the first part was open to traffic later

that year, the full 8-mile stretch had its official grand opening on December 30, 1940 [...] A freeway is defined as a multiple-lane divided highway, with access and exits provided by clover-leaf-style ramps on intersections like those found on surface streets, with traffic lights or stop signs. And it will not have tolls, hence the 'free' component. Now, if we were talking about the first interstate freeway, that would be a different matter (Pennsylvania, Missouri, and Kansas are all in the running there). But as a state freeway, Arroyo Seco Parkway/Pasadena Freeway has a legitimate claim to be the first.' [238]

4. In our opinion, such a teleological interpretation of Walter's robot's behaviour is unwarranted, as their behaviour is implicit in their design, even if the behaviourist practice of ascribing mental qualities to machines has historically been quite common, with even John McCarthy claiming: 'To ascribe certain beliefs, knowledge, free will, intentions, consciousness, abilities, or wants to a machine or computer program is legitimate when such an ascription expresses the same information about the machine that it expresses about a person.' [231]

5. We note that none of the technology discussed in this book is *fully* autonomous in the strict autopoietic sense of the word, as all technology is engineered for a purpose and hence is incapable of exhibiting genuine teleological behaviour.

6. In this compact sense, Tsugawa's achievement is akin to David Horne's subsequent 1 kB chess program, sold for the Sinclair ZX81 computer in the UK in 1982. Although you wouldn't fancy 1K ZX chess's chances against IBM's Deep Blue, just getting a playable chess program to fit into 1 kB remains a significant intellectual achievement in itself (cf. 1 kB ZX chess by Sinclair Research (ZX81): https://tinyurl.com/5fnvc7b4). For comparison, a trivial 'Hello World' program compiled using MinGW-w64 (GCC C++) on Win 10 was recently reported by Stack Overflow to compile to an executable file size over 180 kB (cf. Stack Overflow: https://tinyurl.com/ypapfd24)—such is progress.

7. Jitendra Malik, a computer vision and AV pioneer from the University of California, Berkeley, calls Dickmanns 'the man who actually invented self-driving cars', and points out that computers were a million times slower at this time. [221]

8. PROMETHEUS: 'PROgraMme for a European Traffic of Highest Efficiency and Unprecedented Safety'. [352]

9. In 1987, Dickmanns first demonstrated autonomous driving using Dynamic Vision on the Autobahn, advocating for camera-based systems over buried cables or lidar. Decades later emphasized he emphasized knowledge representation and 4D modelling for predictions, expilictly drawing on Jakob Hohwy's theory of the brain as a prediction-error-minimizing system. Indeed, his work connects to both Hohwy's predictive processing framework, [147] and Rosen's anticipatory systems theory, [284] which, despite different origins in cognitive neuroscience and theoretical biology, respectively, share core principles about internal models and future-oriented cognition.

10. More accurately, ALVINN augmented purely visual information, digitized as a 30x32 video input retina, with a 32x8 array of range finder data, e.g. from a lidar unit.

11. It is essential to note that the system only applied computer vision for steering, i.e. lateral control. The much more difficult task of leveraging AI for lateral and longitudinal control was only achieved by Ernst Dickmanns' response to *no hands across America*, which was the VaMP test-run Munich–Odense of November in the same year. Cf. Long distance drive VaMP 1995, web resource by Ernst Dickmanns on his website: https://tinyurl.com/5cdb3zvk

12. However, this applies for better or for worse, i.e. there are cases where imitation of human-like driving is suspected of causing severe problems. Industry veteran Brad Templeton reports: 'There is suspicion that when a Cruise car made a left turn from the middle lane of a San Francisco street, crossing in front of a backhoe which hit it, this might have been decided by a machine learning planner which had seen too many humans pull stunts like this.' [338]

13. Encyclopedia Britannica: Ancient Egypt. '*to blot out a person's name was to destroy that individual for all eternity, to eliminate him from the historical record*'. [99]

2 Computing Machinery and Intelligence

Historically, legally safe driving has required a responsible human to be in control of the vehicle. However, as we noted in Section 1.1, legislation has recently been emerging globally, permitting machines to take control of vehicles using artificial intelligence control systems designed to match human driving performance. These systems fundamentally rely on computation, but what, *precisely*, is entailed by this notion?

2.1 WHAT IS COMPUTATION?

With computers now ubiquitous in all aspects of 21st-century life, from phones and desktop devices to washing machines and lawnmowers, the question 'What is computation?' may initially seem trivial. However, as we will subsequently unfold, this is far too facile a view.

As recently as 2013, as part of its annual convention, the AISB—the world's longest-established society for the study of artificial intelligence and the simulation of behaviour—was obliged to host a symposium entitled *The Scandal of Computation—What is Computation?* [37]

It transpires that a formal notion of computation, one that encompasses all its forms and varieties, is *extremely* hard to pin down. One recent attempt at proving such a general description entailed deploying Yuri Gurevich's *abstract state machines* as a formal constraint language, and was co-presented at the AISB 2013 meeting by one of the authors of this volume. [331]

Nonetheless, as a first approximation, we can think of computing as the process of performing *calculations*, or processing *information*, by following a set of discrete *rules* or continuous operations, using either discrete (digital) or continuous (analogue) *processes and data.*[1] The goal is to transform inputs, typically via physical or electronic systems, into outputs that perform practical tasks, e.g. perform arithmetic, control engine speeds, drive cars, etc. At the very least, this rough definition allows us to tease apart two central forms of computation: *analogue* and *digital*.

Analogue computation utilises continuous physical phenomena—such as voltage, bodily motion, or fluid flow—to represent values and the behaviour of natural processes operating on these phenomena to solve problems. Digital computation, on the other hand, processes discrete data, typically encoded using binary numbers (0s and 1s), and performs specific operations on the data step-by-step in a logical, precise manner.

Thus, a fundamental difference between the two forms of computation is that analogue systems process *continuous* signals embedded in *time*, while digital systems process *discrete* signals embedded in *sequence*.

DOI: 10.1201/9781003561460-3

2.1.1 ANALOGUE COMPUTATION

One of the most ancient and most complex analogue computing devices was the Antikythera mechanism (2nd century BCE). This hand-powered orrery is the oldest known example of an analogue computing device and was used to predict astronomical positions and eclipses decades in advance. Subsequent notable examples of analogue computers include:

1. **The slide rule (17th century)**: A mechanical analogue device used for multiplication, division, and other functions such as roots and logarithms. It operates by using logarithmic scales, where distances on the ruler correspond to the logarithms of numbers. By sliding the central part (the slider) against a fixed scale, users can quickly add or subtract logarithms, effectively multiplying or dividing the numbers. The slide rule offers an approximate yet efficient method for handling complex calculations without digital precision. It was a popular tool for engineers and scientists until digital calculators supplanted it in the 1970s.
2. **The Watt governor (18th century)**: The Watt centrifugal governor, invented by *James Watt*, was used to regulate the speed of a steam engine under load. It consisted of two metal balls connected to a shaft that rotates at engine speed. The faster the engine runs, the greater the centrifugal force of the balls and the greater the angle they make with the vertical. Completing the feedback loop, this angle controls the amount of fuel supplied to the engine. By adjusting the energy supply to the engine, the governor can regulate the engine's power to maintain a constant engine speed.
3. **The differential analyser (1927)**: Invented by Vannevar Bush, this was an analogue computer designed to solve differential equations by simulating the continuous physical processes they represent. It operates using interconnected mechanical components such as rotating discs, wheels, and gears. These components perform mathematical operations, such as integration, by translating the motion of one part (representing a variable) into the motion of another. The system mimics the behaviour of dynamic systems described by differential equations, making it capable of solving complex, real-world problems, such as trajectory calculations or engineering simulations. It was a crucial computational tool before the advent of digital computers.
4. **The Moniac (1949)**: This water-driven analogue computer, also known as the Phillips Hydraulic Computer, was used to simulate economic systems, modelling the flow of money through pipes and tanks to represent variables like spending and saving. It was used in universities and government institutions in the UK, the US, New Zealand, and elsewhere to help students and policymakers visualize economic systems.
5. **Flight simulators (1960s-1980s)**: Some early flight simulators, for example, the *Link Trainer* and *Curtiss-Wright simulators*, used analogue computers to model flight dynamics and train pilots. These machines relied

on continuous systems to replicate real-world aircraft aerodynamics in real time.

However, although analogue computation played a major part in some of the significant scientific and technical achievements of the twentieth century, the technology is now more rarely deployed and virtually never in AVs, albeit echoes still resonate in digital mirrors of the Watt Governor used by the feedback control systems typically deployed in AVs to regulate engine speed under load.[2]

2.1.2 THE DIGITAL COMPUTER

Historically, the term *computer* referred to a person performing calculations, while *computation* described the human act of calculating. This understanding underwent a profound transformation through the work of Alonzo Church on the *lambda calculus*, [66] and Alan Turing on *Turing machines*, [349] which formalized the concept of an *effective method*: a procedure that, through a finite sequence of simple, and unambiguous steps, is *guaranteed* to terminate and produce *correct* result for the class of problems it is designed to solve.

These developments shaped the modern conception of *computer algorithms* as effective methods for computing functions and defined *effective calculability* as the execution of discrete processing steps by automated systems, to generate outputs in finite time.[3]

In his seminal 1936 paper: *On Computable Numbers with an Application to the Entscheidungsproblem*, British polymath Alan Turing developed an abstract model detailing how mathematicians [human computers] calculated using an effective method—a model of computation now known as the *Turing machine*. [349]

In the Turing machine, a linear tape storing a series of discrete symbols selected from a finite alphabet serves as both the input for a computational problem and as the workspace for subsequent computation. A simple *finite state automaton* (FSA) performs each step of the calculation.[4] In this way, every [Turing] computable problem with input x is formalized by a *specific* Turing machine **P (x)**. If the problem requires more than one input, an encoding function, such as prime decomposition, can be used to uniquely encode the value of multiple parameters onto the tape.[5]

Turing subsequently proved a more general device, known as the *universal Turing machine*, **U**, to be capable of decoding an arbitrary set of computational instructions encoded on the tape, such that, given the encoding of a program P and its input x suitably defined on the tape as input, the symbols left on the tape after the termination of the universal Turing machine **U (P, x)**, is identical to that which would have resulted from the operation of Turing machine **P (x)**; a process analogous to a modern PC running an arbitrary program loaded into its memory. In the universal Turing machine, the abstract concept of modern *digital computation* had arrived.

During the 1940s, Stanislaw Ulam and John von Neumann developed an alternative abstract computing machine called a *cellular automaton*.[6] This is another discrete model of computation, but one that consists of a grid of cells, each in one of a finite number of states. The grid can be of any finite number of dimensions.

Each cell in the grid has a defined neighbourhood. Starting from an initial state at time $t = 0$, new generations are created by advancing t by 1. A fixed rule determines each cell's new state based on its current state and the states of its neighbours. Typically, this rule is uniform across all cells, unchanging over time, and applied in parallel across the entire grid.[7]

Subsequently, the British mathematician and computer scientist Stephen Wolfram proved that certain forms of cellular automata are *computationally universal*, i.e. they are *Turing complete* and are capable of simulating *any* Turing machine. [385]

Furthermore, like the Turing machine, cellular automata do not intrinsically embed a notion of time; instead, they function via sequences of operations, in this case, executed in parallel across the grid of states. To see the difference, one could, for example, randomly vary the physical duration of each step in the sequence, and the computational function(s) computed by both machines would remain the same. To put this more prosaically, one could never [soft] boil an egg with a Turing machine or cellular automaton without first locking the duration of each time step to a reliable external clock.

The invention of the digital computer was a crucial milestone in the subsequent development of autonomous vehicles. Von Neumann's 1945 seminal report *First Draft of a Report on the EDVAC*, [363] proved to be a foundational work on digital computation. In it, von Neumann proposed a computational system with an independent memory module storing both data and machine instructions (*program code*). This is now known as a von Neumann, or Princeton computer (as opposed to a Harvard architecture machine, which employs two separate memories: one for data and one for instructions).

Critically, in the von Neumann architecture, data and instructions are moved over the same hardware (*bus*); hence, the CPU cannot read [or write] data to [or from] memory at the same time as it reads a program instruction. Inevitably, this use of a shared memory bus results in limited memory throughput (*bandwidth*), a phenomenon known as the von Neumann bottleneck, which we will revisit when we discuss *massively parallel processing* in Section 9.1.

Nevertheless, most computers built since the 1950s remain based on the von Neumann architecture. This is the architecture that directly spurred the subsequent development of *integrated circuits* (ICs); tiny, solid-state devices that lie at the heart of modern digital computers. And for these devices, we have observed performance roughly doubling every two years—a phenomenon described by what has become known as Moore's law, after the Intel engineer, Gordon Moore, who first observed it in 1965. Indeed, this tendency—the performance of ICs to improve exponentially—is the fundamental reason behind the incredibly rapid development and widespread application of *digital* technologies over the last seventy years.

2.2 WHAT IS INTELLIGENCE?

Currently, driving a vehicle in any context that a human can requires human *intelligence*. This is, as we said in the beginning, due to the open-ended nature of the real world. Hence, it is crucial to explore precisely what human intelligence *is*, as engineers must computationally replicate [at least some key aspects of] it to build useful autonomous vehicles.

For many years, engineers working on AVs have been preoccupied with mainly practical issues: how to prevent the vehicle from veering off the path, how to avoid shadows or bushes from confusing the vision systems, how best to adjust the driving speed to the environment, etc. They did not need to ask themselves more abstract and fundamental questions, such as: What is intelligence?

However, as efforts to extend the scope of AVs progressed and as AI became more central to this, such abstract questions inevitably arose: What distinguishes human intelligence from artificial intelligence? How can machines be designed to cope with challenges in a way that is as flexible as human reasoning?

The word *intelligence* is derived from the Latin verb *intelligere*, which means to comprehend or perceive. Historically, the term intelligence has been identified with the *relating* ability of the mind with regard to sense data. Indeed, as long ago as 1781, Immanuel Kant identified the '*transcendental unity of apperception*' as the junction at which the self and the world come together and intelligent action as the relating ability of the mind that '*transforms raw sense data into a coherent set of objects and concepts*'.[8] [174] Similarly, in the 1920s, the Gestalt school of psychology identified intelligence and problem-solving with insight and the relating activity of mind, specifically, as an organized, coherent whole, whose parts are determined by laws intrinsic to the whole, rather than merely randomly juxtaposed.[9]

Nonetheless, it is essential to note that, in contrast to the mathematical definition of, say, a triangle, there is no universally agreed-upon definition of intelligence. [61] Instead, we have numerous general definitions of intelligence and multiple benchmarks. General definitions attempt to describe intelligence broadly, while benchmarks link intelligence to the ability to solve specific problems that, when solved by humans, are typically considered to require intelligence.

As we shall show in the following sections, computers have now successfully tackled several core benchmark tasks once thought to be paradigmatic of human intelligence. This raises an important question: Does the fact that computers can now perform these tasks mean that we have already achieved artificial intelligence?

While this question sparks heated debate, it is clear that AI systems' performance in benchmark tests has consistently pushed the boundaries of what were once considered core characteristics of intelligence—a notion called *AI effect* and credited to John McCarthy.[10] And, as many in the field continue to observe trenchantly, this trend persists to this day—every new benchmark smashed seems only to move the goalposts.

2.2.1 BENCHMARK 1. THE ABILITY TO DO IQ TESTS

In 1964, while a PhD student at MIT with Marvin Minsky, Thomas Evans developed software to solve geometric analogy problems, such as those frequently encountered in intelligence quotient (IQ) tests. [101] For example, each member of this class of problems typically consists of labelled line drawings, and the task to be performed is to complete an analogy, e.g. *Figure A is to Figure B as Figure C is to which of the given answer figures?*

As Marvin Minsky observed:

> 'The particularly attractive feature of this kind of problem as a test of machine intelligence is that it has no uniquely correct answer. Indeed, performance on such tests is not graded by any known rule but is judged on the basis of the choices of highly intelligent people on whom the test is tried.' [101]

Indeed, precisely because of the level of difficulty required, such tests were considered to require considerable intelligence to solve and were therefore widely used in college admission examinations across the USA.

2.2.2 BENCHMARK 2. THE ABILITY TO DEDUCE COMPLEX MATHEMATICAL PROOF

Newell and Simon's Logic Theorist program demonstrated strength in this task at the Dartmouth conference in 1956. [261] In operation, the Logic Theorist navigated a *search tree* (see *Problem solving with trees*, pp. 35) with the initial hypothesis as its root. Each branch of this tree represented a deduction based on logical rules. The goal—the proposition the program aimed to prove—existed somewhere within this tree structure. The proof consisted of the path from root to goal, comprising a series of statements. Each of these statements was logically deduced from the previous one, forming a chain of reasoning that led from the original hypothesis to the proposition being proven. In this manner, the Logic Theorist successfully proved 38 of the first 52 theorems in Chapter 2 of Russell and Whitehead's magnum opus, Principia Mathematica.

Furthermore, the Logic Theorist's proof of Theorem 2.85 was noted to be more elegant than the original proof by Russell and Whitehead. Indeed, when Simon presented the new proof to Russell, he '*responded with delight*'. [235] This marked the first instance of a computer proving abstract statements—32 of the 58 mathematical theorems in *Principia Mathematica* [375]—rather than merely performing numeric calculations.

Subsequently, Newell and Simon attempted to publish the new proof in *The Journal of Symbolic Logic*, with the Logic Theorist credited as one of the authors. However, the paper was rejected on the grounds that a new proof of an elementary mathematical theorem was not notable, with the editors overlooking the sensation that one of the authors was a computer program. In 1957, the core ideas behind the Logic Theorist were refined into a more generic problem-solving program, which they called the General Problem Solver (GPS).

In contrast to the earlier Logic Theorist project, the GPS employed *means-ends analysis* and encapsulated a psychological theory of human problem-solving. It was developed using an experimental protocol to characterize the behaviour of a human test-subject when solving a logic problem. Unlike task-specific algorithms, GPS was a general-purpose system capable of addressing a wide range of structured problems, provided they could be encoded symbolically.

In its development, Newell, Simon, and Shaw showed how GPS, a program defined by a recursive structure of goals, subgoals, and means-ends analysis, produced the same results as human subjects following the experimental protocol. For example, when GPS was tested using a specified experimental protocol, such as solving logical puzzles, mathematical problems, or symbolic tasks, it typically followed problem-solving steps that closely resembled those used by humans.

However, while GPS could produce similar results to human subjects in structured domains, notable differences in performance emerged when problems became less well-defined or when intuition, creativity, or contextual understanding were required. [262]

2.2.3 BENCHMARK 3. THE ABILITY TO ANSWER SPECIALIZED QUESTIONS ABOUT A COMPLEX KNOWLEDGE DOMAIN

With the publication, in 1971, by Feigenbaum, Buchanan and Lederberg of *On Generality and Problem Solving: a case study using the DENDRAL program*, [102] a computer demonstrated the ability to answer very detailed and specialized questions. Heuristic DENDRAL was an expert system program written to solve problems of inductive inference in organic chemistry. It was a key development in the timeline of symbolic AI, as it marks a fundamental departure from Newell and Simon's goal of *general* problem-solving, moving towards what became known as *knowledge-based solutions* (aka *expert systems*).

This shift of paradigm, that *knowledge is power*, was signposted and explicated in the 1971 paper; as Feigenbaum trenchantly remarked: 'Knowledge, not tricks or metaphysical insight, is what makes the program effective—and that itself is an insight of general import'. [102]

2.2.4 BENCHMARK 4. THE ABILITY TO ANSWER VERY GENERAL QUESTIONS ABOUT A SIMPLE KNOWLEDGE DOMAIN

Until Terry Winograd's 1973 paper, *A Procedural Model of Language Understanding*, [383] most research into language attempted to separate it into base components that could be studied independently; components constructed (a) primarily as a syntactic program (Kuno, 1965) [192]; (b) a model of semantic connections (Schank, 1971), [299] or (c) as an attempt to model the memory structures (Quillian, 1967). [277]

Winograd's critical insight, which mirrors that of Roy Harris's *integrationist linguistics*, was that prior systems had attempted to address the problems of meaning by focusing almost exclusively on the understanding of single sentences.[11] [129] In contrast, virtually no human use of language takes place in such an artificial setting. Humans are always situated in a context, and within that context, we utilize past experiences to help anticipate and interpret what is to come. With the SHRDLU system, Winograd successfully instantiated a computer program that *understands* language in a limited domain by including a model of the subject being talked about and a context of discourse—an elementary knowledge domain, a *Blocks World*—about which, the questioner was able to pose very general questions.

2.2.5 BENCHMARK 5. THE ABILITY TO PLAY GAMES OF SKILL

'Chinook', a computer program designed to play *draughts/checkers*, was first developed by scientists at the University of Alberta in Canada. In 1994, Chinook won the world *draughts* championship. Similarly, in 1997, after a six-game match, Deep Blue defeated the reigning world chess champion, Garry Kasparov, by winning three games and drawing one, a victory widely considered a huge milestone in the history of artificial intelligence. [53]

Go is an even more complex game to play than chess, and a game that many thought necessarily required human intuition to play well. For example, in 1965, mathematician I. J. Good wrote:

> 'Go on a computer? In order to program a computer to play a reasonable game of Go, rather than merely a legal game, it is necessary to formalize the principles of good strategy or to design a learning programme. The principles are more qualitative and mysterious than in chess and depend more on judgment. So I think it will be even more difficult to program a computer to play a reasonable game of Go than of chess.'

Nonetheless, in 2016, DeepMind's AlphaGo defeated Lee Sedol, a 9-dan professional, in a no-handicap match. The following year, it triumphed over Ke Jie, who had held the world No. 1 ranking for two consecutive years. [322]

Jeopardy! is an exciting American game show that tests the knowledge and language skills of its contestants. Unlike other game shows where questions are asked, in Jeopardy!, contestants are given clues in the form of answers and must come up with the appropriate question. It may seem like a simple game, but Jeopardy! poses unique challenges to computers compared to other games, such as chess, Go, and draughts. However, in 2011, a team of scientists led by David Ferruci at IBM created an AI that competed against previous *Jeopardy!* champions and won $1 million in a televised match. [103]

Historically, AI has been most successfully applied to games that have deterministic rules and known strategies. Games like *chess*, *Go*, and *draughts* are ideal, since they are fully observable, turn-based, and with clear rules and strategies that can be evaluated using tree-search algorithms (see page 35).

Conversely, games like *StarCraft*—a popular military science fiction computer game developed by Chris Metzen and James Phinney—pose a much more significant challenge, as players have over 1,026 possible moves and partial observability; indeed, players have less information about their opponents than in poker. In such types of games, cooperation and the ability to establish trust through dialogue are essential. Hence, the game requires significant (inter-player) communication skills. Even so, AI has made significant strides in playing complex games like *StarCraft*,

[362] with AlphaStar, developed by DeepMind, achieving Grandmaster rank, and placing it among the top 0.2% of human players on the ranked ladder. [361]

Research into this type of subtle, interpersonal gameplay continues at a rapid pace. Diplomacy is a strategic board game that requires players to build alliances and negotiate with each other in order to win. As so many nuanced and paradigmatically human skills are centrally embedded within the game, it is a task previously identified by one of the authors of this book as a particularly robust challenge for machine intelligence. [34] However, in 2023, Meta published the results of its work on developing an AI system that can play *Diplomacy*, if not yet at grandmaster level, still surprisingly well. [20]

Even so, we contend that such benchmarks merely test *aspects* of intelligent action, not intelligence *per se*. Thus, although AI has dramatically pushed the boundaries of what was once considered the purview of humans, the ability to autonomously drive a car in all environments remains elusive. Or, as Yann LeCun remarked:

> 'Still, our best M[achine]L[earning] systems are still very far from matching human reliability in real-world tasks such as driving, even after being fed with enormous amounts of supervisory data from human experts, after going through millions of Reinforcement Learning trials in virtual environments, and after engineers have hardwired hundreds of behaviours into them.' [198]

Although some researchers have questioned whether some modern AI systems, such as ChatGPT, have already effectively passed the Turing Test, [28] [166] no one can yet claim that an AV can operate seamlessly in the unbounded open world. With self-driving cars, AI is applied to a world of infinite possibilities, uncertainty, and imperfect information. To do this successfully, we argue, requires intelligence *per se*.

2.2.6 INTELLIGENCE *PER SE*

To understand how research defines the term *intelligence* and derive a comprehensive definition from it, a 2019 scientific paper analysed the most critical works on the subject. They found 70 different definitions, which they synthesized into one, as derived from a 2007 analysis by Legg and Hutter [158]):

> '*Intelligence measures an agent's ability to achieve goals in a wide range of environments.*' [207]

Intelligence in machines, much like in humans, is about making the right choices to achieve specific goals. This concept has been a critical challenge for engineers developing self-driving cars who view intelligence *pragmatically* as the ability to act appropriately within a given environment to reach defined objectives.

Stuart Russell, a prominent AI researcher, put it simply in 2018:

> 'An entity is intelligent to the extent that it does the right thing, meaning that its actions are expected to achieve its objectives.' [105]

This idea was not new, as it echoed the thoughts of Jean Piaget, a famous developmental psychologist, who, decades before, suggested that intelligence centrally concerns *adapting to one's environment.* [380]

Piaget's earlier work in 1950 proposed that intelligence is not a single trait, but rather exists to some degree in all animals, enhancing their ability to interact with their surroundings in complex ways. Building on this, researcher Philip E. Vernon noted in 1960 that intelligence relates to a person's ability to process information flexibly and respond to their environment in a complex manner. [360]

The most accepted explanation—according to a survey of 400 leading AI professionals in 2018 [247] [160]—is a working definition by leading AGI researcher Pei Wang,[12] that fundamentally emphasizes *real-world constraints*, including insufficient knowledge, limited processing capacity and coping with novelty:

> 'The essence of intelligence is the principle of adapting to the environment while working with insufficient knowledge and resources. Accordingly, an intelligent system should rely on finite processing capacity, work in real-time, be open to unexpected tasks, and learn from experience.' [160]

The aspect of dealing with novelty is also highlighted by François Chollet, the inventor of the ARC-AGI benchmark,[13] however the definitional introduction of time, processing, and energy constraints relates explicitly to our purpose of studying real-world agency. Such constraints are critical for autonomous robots, but not for virtual chatbots. In terms of such constraints, the goalposts for AGI have remained steadfast since the beginning.

Thus, to exhibit intelligence *per se*, a machine needs several key components: goals, the ability to perceive its environment in real-time, decision-making capabilities under uncertainty, motor control, intention, continual learning, and eventually, language and reasoning skills. As Stuart Russell explains:

> 'We're talking about the ability to make decisions, to plan, and to problem-solve. We're talking about the ability to communicate, and so natural language understanding also becomes extremely important to AI.' [105]

Nonetheless, it is essential to note that real-world problems often do not have clear-cut, objectively correct answers. Therefore, intelligence also involves organizing oneself when faced with uncertainty and choosing options that will approximate or (iteratively) lead to a desired outcome *from the perspective of the decision-maker*. It entails not just pursuing goals but also actively setting and refining goals. In this context, having *preferences* becomes a crucial prerequisite for intelligent, *teleological* behaviour. Furthermore, although intelligence *per se* is not necessarily tied to the brain or nervous system, it is [to date] tied to animate matter, as Evan Thompson elegantly foregrounded in *Mind in Life: Biology, Phenomenology, and the Sciences of Mind*, an exegesis on these themes. [341]

For example, plants, living organisms without a brain, exhibit remarkable adaptive behaviours that some researchers refer to as *plant intelligence*. These include the ability to communicate through chemical signals, demonstrate forms of memory and learning, solve problems like navigating obstacles, perceive and respond to various environmental stimuli, adapt to changes, defend against threats, and maintain circadian rhythms. Some plants can even distinguish between self and non-self roots or engage in parasitic targeting.

These sophisticated behaviours challenge our understanding of intelligence and adaptation in the plant kingdom, even if the term *plant intelligence* remains controversial in scientific circles. Reminding us of the evolutionary origin of intelligence, it demonstrates how intelligence is profoundly situational and context-dependent, whether embodied in a brain or not.

Intelligence is not an abstract one-dimensional quality that can be scaled up independently of context by simply adding resources. The environment, shaped by circumstances, is an integral part of a system. This is why there is no hierarchy of species or varying levels of intelligence; instead, there are different implementations based on varying situational requirements. Ultimately, the success of any expression of intelligence is measured by one currency: successful adaptation within a specific survival niche.[14]

Nonetheless, even though, as Legg and Hutter also remark, a formal definition of intelligence remains elusive, [206] it is clear from Piaget and Vernon that being able to act appropriately in an open environment and adapt to change is a critical aspect of intelligent behaviour. Another key aspect is *learning*: the autonomous or self-directed acquisition of desired behaviour(s) from [sense] data.

These were precisely the kind of problems that the pioneers of autonomous driving needed to solve too—but practically and step-wise—constantly adapting to the current objective, not as an abstract research programme into artificial intelligence; albeit, without realizing it, they were part of precisely that.[15]

To better understand the overarching questions to which AV engineers were unwitting contributors, let's examine the evolution of artificial intelligence and the questions it raises. In later chapters, we will build on this knowledge to evaluate the prospects of AI-based, Level–5 autonomous driving technology.

2.3 ON MECHANISING INTELLIGENCE

The idea of creating an artificial—*mechanical*—mind isn't new; its precursors date back to ancient times. Although the term *artificial intelligence* was only coined in 1955 by John McCarthy—in his research proposal for the Dartmouth conference (see page 36)—the fundamental concept of creating machines that exhibit intelligent behaviour can be traced much earlier. For instance, in ancient Greece, engineering was sufficiently advanced to make artefacts, such as clepsydra (water clocks) and automata, that exhibited a single, predetermined, human-like behaviour.

2.3.1 ANCIENT GHOSTS IN THE MACHINE

'Ctesibius of Alexandria', a Greek inventor (285–222 BCE), significantly advanced *Clepsydra* technology. His devices were far more complex than earlier models, featuring intricate displays of time and complex mechanisms to regulate water flow precisely. The displays ranged from simple bells and gongs to elaborate automatons and even astronomical models. Furthermore, his most advanced mechanisms incorporated a groundbreaking *control system* to maintain a constant water level in the reservoir and ensure considerably more accurate timekeeping. This innovative approach marked the earliest known example of *feedback control*, a fundamental principle in cybernetics. Astonishingly, for over 1,800 years, until usurped by the advent of the pendulum clock in 1656, Ctesibius' clepsydra reigned as the world's most accurate timekeeping device.

However, it was not until the 19th century, at the dawn of the Victorian era in England, that Charles Babbage and Ada Lovelace laid the technical foundations for both digital computing and what we now consider artificial intelligence. These foundations eventually facilitated the construction of machines capable of even more complex 'intelligent' actions.

For his inventions of the *Difference Engine* as well as the *Analytical Engine*, Charles Babbage is widely regarded as the father of the computer. The former was a machine designed to automatically perform some of the complex calculations of ballistic trajectories required by the military. Although the latter, the Analytical Engine, the precursor of the modern computing engine, was never completed, it is generally considered that Babbage's ideas forged a foundation for contemporary digital computing. In 1842, Luigi Federico Menabrea published a French description of the Analytical Engine based on Babbage's 1840 Turin lectures. [240]

Ada Lovelace first met Charles Babbage on 5 June 1833 at a party in London. Lovelace, then 17, was already deeply interested in mathematics and science, while Babbage, aged 41, was a well-established mathematician. Mary Somerville, a Scottish scientist and Ada's tutor, introduced them, seeing the potential for an intellectual connection. Their encounter was characterized by mutual respect and intellectual curiosity. Lovelace was fascinated by Babbage's innovative work on computing machines, particularly the Analytical Engine. Babbage, in turn, was struck by Lovelace's remarkable mathematical insight, which was uncommon among women of the time.

Subsequently, in 1843, Ada translated Menabrea's analysis of the Analytical Engine into English, adding extensive notes of her own. [241] Her additions included a method to calculate the 'Numbers of Bernoulli' using the machine, now often considered the first complete computer program (*ibid*). This contribution eventually led to Lovelace being recognized as the world's first *computer programmer*. Later, she identified and corrected the first-ever *computer bug* in that code. More importantly for our narrative, Lovelace also recognized the potential for computers to go way beyond mere calculation and be used for more creative purposes.[16] Indeed, in a remarkable premonition of the latest *generative AI* tools, Ada specifically wrote about the possibility that computers 'might compose ... scientific pieces of music of any degree of complexity or extent'. [16] Hence, it is now widely agreed that Lovelace *anticipated* the development of modern artificial intelligence.

Babbage and Lovelace's work helped establish the basic principles of computing and programming, which are essential for AI systems. The Analytical Engine's design incorporated many of the same components used in modern computers, including a central processing unit (CPU), memory, and input-output devices. The Analytical Engine's instructions, coded on punched cards, were also seen as a precursor to modern computer programming. In this way, their work not only helped advance technology but also transformed the way we think about machines and their potential.

In the following century, in 1910, the Spanish engineer Leonardo Torres y Quevedo embarked on a groundbreaking project that resulted in the construction of a physical chess automaton, which he christened *El Ajedrecista* (The Chessplayer). [346] This machine boasted the remarkable ability to autonomously engage in a *King and Rook* endgame against a king from any initial position.

Initially, mechanical arms manoeuvred the Chessplayer's pieces, but by the year 1920, electromagnets installed beneath the board facilitated this function. Hence, this invention may be considered the pioneer of modern, interactive computer games, and it evoked unprecedented enthusiasm upon its public debut at the prestigious University of Paris in the year 1914. Nonetheless, the world had to wait until 1948 for Alan Turing and David Champernowne to code *Turochamp*, the first program that could play a full game of chess, and it is this achievement that we choose to posit as the birth of modern artificial intelligence.[17]

From these early roots, an exciting new and heterogeneous field with different goals, assumptions, and working methods developed; a field which included cybernetics, systems theory, automata studies, automation control theory, and complex information processing; the field of artificial intelligence.[18]

2.3.2 CYBERNETICS OR CONTROL AND COMMUNICATION IN THE ANIMAL AND THE MACHINE

The field of cybernetics was initiated by the American mathematician and philosopher Norbert Wiener. [378] During the war, Wiener worked on the automatic targeting of anti-aircraft missiles, incorporating the pilot's behaviour—knowing he was being followed—into a feedback loop for more accurate predictions.

In 1943, Wiener met with computer pioneer John von Neumann and other engineers and neuroscientists for an interdisciplinary exchange on *circular causal processes* and the *similarities between the brain and computers*. In 1947, he and other scientists coined the term *cybernetics* to describe this line of research, after a paradigmatic example of circular causal feedback—that of steering a ship; indeed, the term cybernetics stems from the Greek *Kybernetes* meaning steersman.[19] Wiener published its central manifesto, '*Cybernetics or Control and Communication in the Animal and the Machine*', [378] just one year later; a work strongly intertwined with related considerations, e.g. from Claude Shannon and Marshall McLuhan, on information theory and systems theory, respectively.

The Macy Conferences were a series of influential events, with the cybernetics strand playing a pivotal role in the early development of AI. Organized by Frank

Fremont-Smith at the Josiah Macy Jr. Foundation, these meetings were held in New York between 1946 and 1953. While the Macy Conferences covered a range of themes, the *Macy Cybernetics Conferences*, driven by Lawrence K. Frank and Fremont-Smith, focused on crucial topics that would later shape AI, including self-regulating and teleological systems, simulated neural networks, machine learning, object perception, and the derivation of ethics from science and data. Chaired by Warren McCulloch, a significant figure in artificial intelligence and neuroscience, these discussions had a lasting impact on the field.

2.3.2.1 The McCulloch–Pitts neuron model

The earliest cybernetic work in the emerging field of neural computing sought to understand, model, and emulate neurological functions and learning in animal brains. This foundation was laid by neurophysiologist Warren McCulloch and young self-taught mathematician Walter Pitts in their landmark 1943 paper entitled, '*A logical calculus of the ideas immanent in nervous activity*'. [236] McCulloch was already an established professor. At the same time, Pitts, an autodidact who had taught himself logic and mathematics as a child, was an unlikely prodigy—a runaway from a family that could not appreciate his genius. McCulloch and his wife took him in, and the two scientists developed their theories during late-night discussions.

Despite his youth, Pitts was already an accomplished logician, a protégé of the eminent positivist Rudolf Carnap, and merely twenty when the groundbreaking paper with McCulloch was published. Eight years earlier, at the age of twelve, he had read Whitehead and Russell's *Principia Mathematica* in just three days in a library, and subsequently wrote to Bertrand Russell, identifying issues with the work. Russell was impressed, invited the young reader to his Chicago lectures, and maintained their correspondence.

For his contributions to the paper with McCulloch—the momentous birth of a transformative idea known as *connectionism*—and after five years of unofficial study, the University of Chicago awarded Pitts an *Associate of Arts*; his only earned degree. He later went on to work with Norbert Wiener at MIT, until their acrimonious parting in 1952. From 1946 onwards, Pitts was a key contributor to the Macy Conferences.[20]

In order to describe how the core processing elements of the brain might function, McCulloch and Pitts showed how simple electrical circuits, connecting groups of linear threshold functions, could compute a variety of logical functions. The basic idea was that all logical relations can be represented by combinations of the logical operations: AND, OR, and NOT. In this way, the two researchers demonstrated how networks of such artificial neurons could perform logical calculus and thus implement all the properties of a Turing machine.

In their model, McCulloch and Pitts provided a first, albeit highly simplified, mathematical account of the chemical processes that define neuronal operation and, in so doing, realized that the mathematics that describes the neuron operation exhibited the same type of logic that Shannon deployed in explaining the behaviour of switching circuits: namely, the calculus of propositions.[21]

McCulloch and Pitts realized (a) that neurons can receive positive or negative encouragement to fire, contingent upon the type of their synaptic connections (*excitatory* or *inhibitory*) and (b) that in firing, the neuron has effectively performed a computation; once the effect of the excitatory/inhibitory synapses is taken into account, it is possible to *arithmetically* determine the net effect of incoming patterns of signals innervating each neuron.[22]

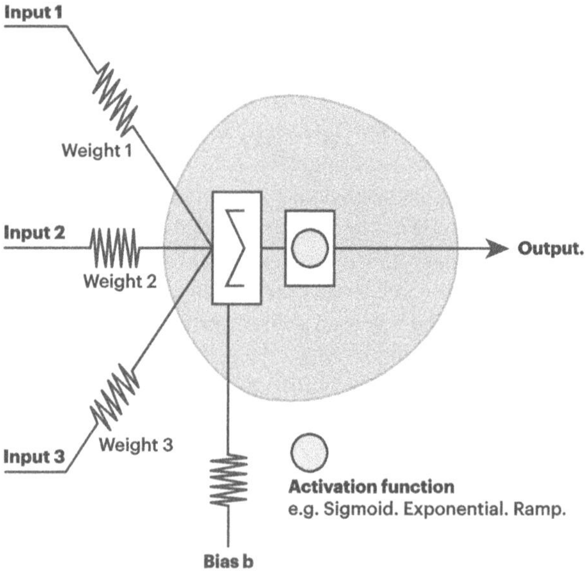

Figure 2.1 The generalized McCulloch–Pitts neuron model. (Image adapted from Bishop, 2016 [33].)

In a subsequent refinement of the basic McCulloch–Pitts (MCP) neuron, cell output is defined by a further [usually non-linear] function of the weighted sum of its input; the neuron's *activation function* (see Figure 2.1).[23]

Critically, McCulloch and Pitts demonstrated that with appropriately chosen synapse polarity, any single input pattern can be recognized by a suitable network of MCP neurons. In other words, any finite logical expression can be implemented using an appropriate network of McCulloch–Pitts neurons:

> 'It is easily shown: First, that every net, if furnished with a tape, scanners connected to afferents, and suitable efferents to perform the necessary motor operations, can compute only such numbers as can a Turing machine; second, that each of the latter numbers can be computed by such a net; and that nets with circles can be computed by such a net, and that nets with circles can compute, without scanners and a tape, some of the numbers the machine can, but no others, and not all of them. This is of interest as affording a psychological justification of the Turing definition of computability and its equivalents, Church's A-definability and Kleene's primitive recursiveness.' [236]

In other words, the McCulloch–Pitts result demonstrated that networks of artificial neurons could be mathematically specified to perform computations of immense complexity and power, and in so doing, opened the door to *connectionism*—a radically new form of problem solving based on the design of neural network architectures and automatic (machine) learning of appropriate network parameters.

2.3.2.2 Symbols and signals

While the McCulloch–Pitts artificial neuron was to serve as a concrete entity on which to scaffold a connectionist approach to building a brain, the symbol proved a far more intangible notion on which to build the alternative: symbolic AI or what John Haugeland dubbed 'good old-fashioned artificial intelligence' (GOFAI). [131] Indeed, as Ernst Cassirer suggested in his 1944 book, '*An Essay on Man*', [56] the *symbol* is not so much a *concrete entity* of human thought, but more a new *dimension* of human reality:

> 'Man has, as it were, discovered a new method of adapting himself to his environment. Between the receptor system and the effector system, which are to be found in all animal species, we find in man a third link, which we may describe as the symbolic system. This new acquisition transforms the whole of human life. As compared with the other animals, man lives not merely in a broader reality; he lives, so to speak, in a new dimension of reality.'

For Cassirer, a symbol has no actual existence as part of the physical world; it has a *meaning*, a lexical semantics. Symbols—in Cassirer's sense of this term—cannot be reduced to mere signals, for symbols and signals belong to two different universes of discourse; signals are operators and part of the physical world of being—e.g. a puff of smoke, a flag, or 5 V in a digital circuit—which can be syntactically combined—e.g. if the signal is digital, by the logical operators AND, OR, NOT, etc. In contrast, symbols are designators and part of the human world of *meaning*.[24]

2.3.3 SYMBOLIC AI

2.3.3.1 Physical symbol systems

In contrast to Cassirer, the so-called *symbolists* conceived the mind (and, for that matter, the computer) as a Physical Symbol System (PSS). [263] Here, the term *physical* is used to denote that the symbol system (i) obeys the laws of physics, is 'realizable by engineered systems made of engineered components', and (ii) is 'not restricted to human symbol systems'.

Thus, contra Cassirer, for the symbolist, a PSS defines a *symbol* syntactically as a set of *physical patterns* (or signals), wherein such patterns, or *tokens*, can be the components of a larger *symbol structure* (or expression). A *symbol structure* is thus composed of tokens related physically, e.g. by adjacency, plus processes capable of operating on such expressions to produce further expressions.

Hence, a PSS is a set of symbol structures mechanically evolving to generate intelligent action; or, as Allen Newell und Herbert A. Simon so elegantly suggested in their 1976 manifesto address to the ACM: '*What is a symbol, that intelligence may use it, and intelligence, that it may use a symbol?*'.

From this insight, Newell and Simon derive their Physical Symbol System Hypothesis (PSSH), [263] an empirical and qualitative law that fundamentally asserts: '*A physical symbol system has the necessary and sufficient means for general intelligent action*':

> 'By *necessary*, we mean that any system that exhibits general intelligence will prove upon analysis to be a physical symbol system; by *sufficient* we mean that any physical symbol system of sufficient size can be organized further to exhibit general intelligence. By *general intelligent action*, we wish to indicate the same scope of intelligence as we see in human action: that in any real situation, behaviour appropriate to the ends of the system and adaptive to the demands of the environment can occur within some limits of speed and complexity.'

The PSSH is an idea philosophically grounded on the work of Frege, Russell and Whitehead and heir to the entire atomistic reductionist tradition in Western philosophy, as Stuart Dreyfus und Hubert Dreyfus elegantly highlighted [95]:

1. From Rene Descartes, understanding is viewed as the forming and manipulation of appropriate mental representations formed from primitive [simple] elements, *naturas simplices* and all phenomena are understood as combinations of these. [85]
2. From Thomas Hobbes, we deduce that these simple elements are formal and are related by purely syntactic—rule-based—operations; hence 'all reasoning is reducible to calculation'—ratiocination. [143]
3. From Gottfried Wilhelm Leibniz, to avoid an infinite regress, there must be at base simple elements which represent things in the world and which are mixed to define complex objects. In other words, there must be 'an alphabet of human thought'. [208]
4. And from (early) Ludwig Wittgenstein's atomistic, picture theory of meaning (as defined in his *Tractatus Logico-Philosophicus*), we see the world as the totality of 'facts, not of things'; wherein *facts* are logically defined by (syntactic) *pictures*, with elements of the pictures being combined in various ways to represent the ways things are combined in the world. [384]

Wittgenstein's Tractatus has been viewed as the culmination of the Western rationalist tradition in philosophy, and Dreyfus and Dreyfus suggest that symbolic AI can be viewed as the quest to find its primitive elements. As they wryly observed: '*Newell and Simon's Physical Symbol Systems Hypothesis effectively turns Wittgenstein's [early] philosophical vision into an empirical claim about the world and bases an entire research programme on it.*'

Furthermore, as Newell and Simon assert that every intelligent system, including humans, is at heart a PSS—an instance of a universal Turing machine—all general intelligent [human] action must be, in principle, realizable by a suitably programmed computer. Thus, if the PSS hypothesis is correct, then Musk's generalized autonomy, *a computer system at least as powerful as the human mind in every aspect*, necessarily follows. This premise of computability is, incidentally, the underlying common feature of the seemingly disparate approaches to AI, since both schools operate on this same basic assumption: the computational theory of mind.

2.3.3.2 Problem solving with symbol structures

Although the Physical Symbol Systems Hypothesis asserts that a symbol system has the necessary and sufficient means for general intelligent action, it does not specify *how* to use a PSS to achieve this goal. Indeed, in the early days of what became known as GOFAI, many programs (such as Newell and Simon's General Problem Solver) were derived simply from observation of human problem-solving behaviour; the common principle that emerged from these observations was that of heuristic search—a rule-of-thumb used to guide the problem-solving process.[25]

In using heuristic search as the PSS engine in an intelligent system, problem solutions are represented as symbol structures, and the PSS exercises intelligence by [heuristically guided] search, i.e. the PSS progressively modifies a symbol structure until it generates a viable problem solution.

To do this, it is necessary to establish (i) a test for a class of successful symbol structures and (ii) a generator of possible solutions. The PSS then simply repeatedly *generates and tests* the symbol structure until it satisfies the test for a valid solution. Sadly, this process does not work *like a dream*, in that we can't simply generate a viable solution in one step, as we cannot know how good any putative symbol structure/solution is until we evaluate/test it. Hence, the process must iterate until an acceptable solution is found, and for many problems this can take a very long time.

2.3.3.3 Problem solving with trees

A tree is an important data structure often deployed in symbolic AI systems to enable systematic, efficient exploration of multiple potential problem solutions; indeed, one such early application of trees in symbolic AI was by the Logic Theorist, as we described on page 23.

The tree structure begins at the *root node*, which represents the initial problem state. From this root, branches extend, each branch corresponding to a set of one or more potential solutions. At the end of each branch is a *leaf node*, defining a specific solution. *Child nodes* between the root and leaves represent intermediate states or decision points in the overall decision-making process. By traversing the tree, one can systematically evaluate potential problem solutions and identify the most promising ones.

However, searching the entire tree can be highly time-consuming and computationally expensive. For example, in chess, the growth of the search tree is influenced by the branching factor, which refers to the average number of legal moves available from any given position. For chess, this average branching factor is approximately

35, although this can vary depending on the stage of the game, e.g. fewer in the endgame, more in middlegame. This means that, on average, a player has around 35 legal moves at their disposal during each turn. For every additional ply (*half-move*), the number of nodes to evaluate multiplies by the branching factor, i.e. depth $n : 35^n$. Thus, at a depth of 10 plies (5 moves for each player), the tree can encompass around 35^{10}, which is approximately 24 billion positions.

Hence, to enhance efficiency, a selection- or tree policy is typically employed. This policy strikes a balance between *exploring* new potential solutions and *exploiting* the most promising ones already discovered. However, the effectiveness of this approach depends on the nature of the problem.

For randomly distributed solutions, a tree search may not outperform a random search across all possibilities. However, in tasks requiring intelligence, where solution distributions follow patterns, an intelligent tree policy can adaptively discover these patterns. This enables informed decisions about which solutions to evaluate next, making the search process more efficient and effective.

2.3.4 THE DARTMOUTH CONFERENCE

As the issues surrounding computation and intelligence became increasingly contentious and the field broader, initial attempts were made at consolidation. In 1955, John McCarthy applied for funding for the Dartmouth conference, during which the various factions were to meet and attempt to draw up an interim balance. In order not to favour any of the different directions, McCarthy needed a neutral title. Hence, the interdisciplinary conference he hosted, with leading scholars in the field—including Marvin Minsky, Nathan Rochester and Claude Shannon—was called 'The Dartmouth Summer Research Project on Artificial Intelligence'. [234] Not everyone was content with this phrasing, but McCarthy's catch-all phrase, *artificial intelligence*, certainly stuck.

Under the aegis that 'every aspect of learning or any other feature of intelligence can in principle be so precisely described that a machine can be made to simulate it', the goal of the aforementioned Dartmouth conference was to exchange ideas on, and align, evolving concepts for thinking machines; ideas which were already quite diverse by this time:

> 'We propose that a 2-month, 10-man study of artificial intelligence be carried out during the summer of 1956 at Dartmouth College in Hanover, New Hampshire. The study is to proceed on the basis of the conjecture that every aspect of learning or any other feature of intelligence can, in principle, be so precisely described that a machine can be made to simulate it. An attempt will be made to find how to make machines use language, form abstractions and concepts, solve kinds of problems now reserved for humans, and improve themselves. We think that a significant advance can be made in one or more of these problems if a carefully selected group of scientists work on it together for a summer.' [234] However, McCarthy later refined his definition of intelligence to be, '... *the computational part of the ability to achieve goals in the world.*' [232]

Nonetheless, the circumstances of the title's emergence already indicated a peculiarity of the emerging field: the lack of a coherent research question, a connecting methodology, and a unifying research programme. James Moore described this in a reflection on the conference as follows:

> 'In some respects, the 1956 summer research project fell short of expectations. The participants came at various times and worked on their own projects, and hence, it was not really a conference in the usual sense. There was no agreement on a general theory of the field, and in particular, on a general theory of learning. The field of AI was launched not by agreement on methodology or choice of problems or general theory, but by the shared vision that computers can be made to perform intelligent tasks.' [250]

As can be seen in Figure (2.2), the methodological pluralism and accompanying lack of a general *theory of AI* persists into the present day; as a field, artificial intelligence remains a loose connection of distinct research agendas, including computer vision, natural language processing, robotics, etc. However, following the taxonomy from Dreyfus and Dreyfus, [95] AI systems can be categorized as belonging to one of two distinct *schools*, contingent on how they conceptually *represent and process knowledge about the world*: either as discrete *symbols* or continuous *signals*.[26]

Continuous signals: Artificial neural network systems typically represent and process knowledge in a *distributed* manner, which mathematically can be efficiently expressed as *tensors*.[27] It is an approach that is focused on the functional properties of neuronal interactions and is best understood within the framework of *psychology* and *cognitive neuroscience*. We might take McCulloch and Pitts' 1943 description of the *Turing equivalence* of computation via neural nets as the launching point for this approach (see page 41). [236]

Discrete symbols: Symbolic AI systems typically represent knowledge by structures of discrete *symbols*; an approach which entails instantiating a formal world model, is philosophically rooted in the *rationalist* and *reductionist* traditions, and is best conceived of under the aegis of *formal logic*. For this, we might choose Newell, Simon, and Shaw's first public demonstration, at Dartmouth conference, of their landmark Logic Theorist as a foundational milestone for this approach (see page 23). [261]

2.3.4.1 Artificial narrow and general intelligence

Although the Dartmouth conference established an early goal of AI—*to simulate every aspect of human learning, or any other feature of human intelligence, by machine*—today, despite the hype around ChatGPT and generative AI, most commercially driven AI activities continue to focus on more *narrowly* specified activities. As a whole, this can be viewed as an attempt to automate as many of the *functions* associated with intelligence as possible, which breaks down into various subdomains, including language, pattern recognition, reasoning, planning, learning, problem-solving, and others.

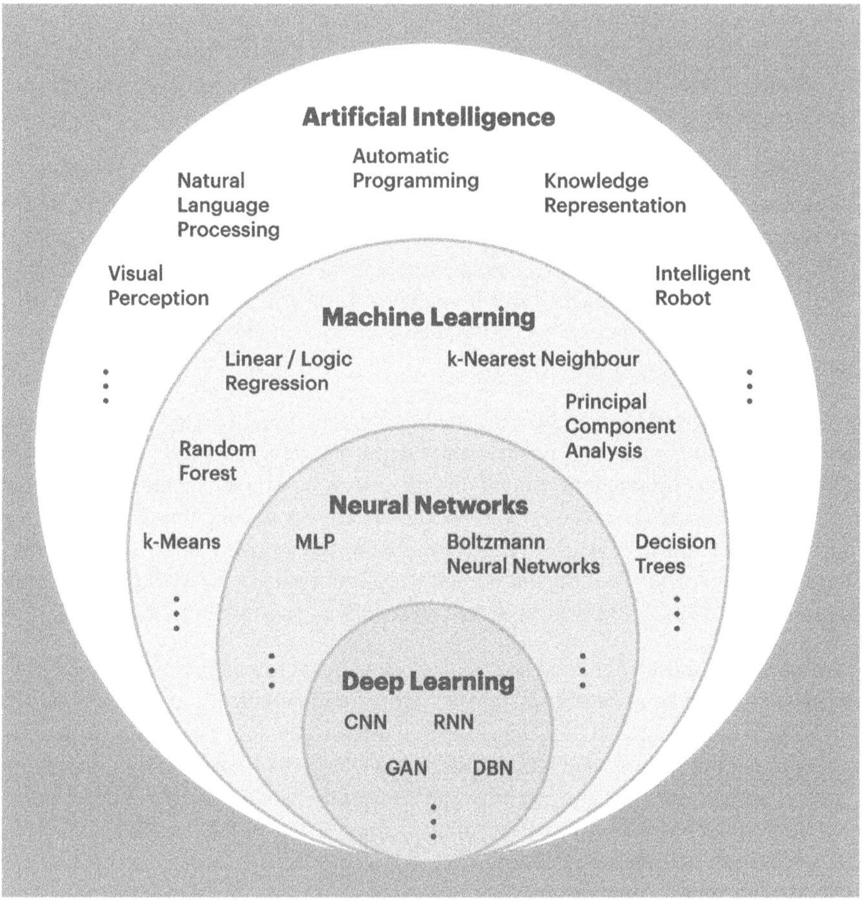

Figure 2.2 The field of artificial intelligence. (Image adapted from Song Li et al., 2021 [210].)

To these ends, two major lines of thinking have evolved about AI:

1. **Strong AI**: these days, often referred to as artificial general intelligence or human-level AI, this approach seeks to build machines that are at least as good as humans in *all* contexts, such that, as the philosopher John Searle famously suggested, the *machine* 'really is *a mind*'. [314]
2. **Weak AI**: also known as artificial narrow intelligence (ANI), this specifically task-based form of AI aims to engineer intelligent systems to perform a set of specified tasks as well as, or better than, humans. Researchers working within ANI are not unduly concerned with understanding how human reasoning actually works, or whether computers genuinely understand. Instead, they focus solely on practical results, and to date, they have been highly successful in delivering such results.

Thus, ANI achieves its practicality by focusing on a specific subset of cognitive abilities, sometimes referred to as special-purpose intelligences—a subset that is gradually improved and extended over time. [282] Such systems process information from specific—and often predefined—data set(s), and their operation is not well defined outside of the task for which they were initially designed. Nevertheless, these systems can be engineered to process large data sets very quickly, often in real-time, which can allow for both a higher pace and deeper information retrieval than humans are typically able to achieve. However, ANI systems fundamentally remain goal-specified and task-based, unable to advance beyond their predefined scope, as they operate under a narrow set of constraints and limitations.

Computer scientists Vladimir G. Ivancevic and Tijana T. Ivancevic describe the terminological question in the following way:

> 'The term *artificial intelligence* would equate to the same concept as what we call *Strong AI* based on the literal meanings of *artificial* and *intelligence*. However, initial research into artificial intelligence was focused on narrow fields, such as pattern recognition and automated scheduling, in hopes that they would eventually allow for an understanding of true intelligence. The term *artificial intelligence* thus came to encompass these narrower fields, the so-called Weak AI, as well as the idea of Strong AI.' [162]

In contrast to Weak AI, Strong AI refers to an artificial system that mimics the human ability to learn in a self-governed way and to apply intelligence to solve open problems, master novel or unknown situations, make judgements under uncertainty, integrate prior knowledge in decision-making, update and revise prior expertise in the light of novel contexts, and perform model-based simulations and counterfactual reasoning, to name but a few. Thus, a putative Strong AI will be able to perform any intellectual task that a human can accomplish, at least as well as, if not better than, a human.

Notes

1. Albeit the notions of calculation (cf. Turing), information (cf. Shannon), rules (cf. Wittgenstein) and data (cf. Floridi) are also very complex ideas to unfold. For our purposes, a common-sense view of these ideas will suffice.

2. See also recent *dynamic systems theories* of mind, [276] wherein Port and van Gelder offer a new theoretical approach to cognitive science, which they suggest may mitigate some of the problems associated with digital computational theories of mind, issues we will explore further in the third volume of this series *The Red Book, Rethinking Autonomy*.

3. The term *algorithm* originates from the name of the great Islamic mathematician Muhammad ibn Musa al-Khwarizmi. It initially referred to place-based arithmetic procedures. His name, rendered as *Algorismus* in late medieval Latin, was used to denote numerical operations. Over time, it evolved into an *algorithm* (analogous to a logarithm), finding its way from French into English.

4. From Wikipedia, '*A Finite-State Machine (FSM) or Finite-State Automaton (FSA, plural: automata), finite automaton, or simply a state machine, is a mathematical model of computation. It is an abstract machine that can be in exactly one of a finite number of states at any given time. The FSM can change from one*

state to another in response to some inputs; the change from one state to another is called a transition. An FSM is defined by a list of its states, its initial state, and the inputs that trigger each transition.'

5. Every positive integer is uniquely decomposable into a product of primes raised to cardinal powers, e.g. $45 = 2^0 \times x^2 \times 5^1$. *Converse*: The product of any set of primes (raised to cardinal powers) is a unique integer.

6. John von Neumann and Stanislaw Ulam developed the concept of a cellular automaton, with notable references in von Neumann's work. The foundational paper detailing this concept, *Theory of Self-Reproducing Automata*, was worked on by Von Neumann in the 1940s. Although Arthur Walter Burks completed and published it posthumously in 1966, Ulam's suggestions influenced the theoretical underpinnings and discussions about cellular automata. [260]

7. Exceptions include stochastic and asynchronous cellular automata.

8. For an overview of Kantian ideas in AI research cf. Hyeongjoo Kim and Dieter Schönecker. [182]

9. An idea which found experimental support in the work of the early cyberneticians. [209] It is interesting to observe that the authors of this paper went on to become extremely influential in artificial intelligence and cognitive science: McCulloch and Pitts emerged as fathers of *connectionism*, and Umberto Maturana [alongside Francisco Varela] went on to be the father of both autopoiesis and enactivist cognitive science.

10. Another reference to this theme appears in Douglas Hofstadter's book *Gödel, Escher, Bach: An Eternal Golden Braid*, where Hofstadter quotes Larry Tesler: 'Intelligence is whatever machines haven't done yet.' [146]

11. That language usage is intrinsically, and without fail, contextual in all of its uses.

12. Derived from Wang's 2008 paper, 'What Do You Mean by "AI"?' [368] Interestingly, in 2019, Wang chose the shorter, but less comprehensive, phrase: 'Intelligence is the capacity of a system to adapt to its environment while operating with insufficient knowledge and resources'. [369] For a deeper dive into these themes, see the *Journal of Artificial General Intelligence* special issue (2020). [248]

13. Chollet writes, 'Intelligence is a cognitive mechanism that you use to adapt to novelty, to make sense of situations you've never seen before. It also works by creating models on the fly of the new situation by combining existing building blocks. Abstract building blocks which were mined from your past experience.' [63]

14. Max Bennett eloquently states in his *history of intelligence*: 'Many niches—in fact, most niches—are better served by smaller and simpler brains (or no brains at all). Big-brained apes are the result of a survival strategy different from that of worms, bacteria, or butterflies. But none are *better*. In the eyes of evolution, the hierarchy has only two rungs: on one, there are those that survived, and on the other, those that did not.' [26] The situational aspect of intelligence earlier prompted François Chollet to question the concept of general intelligence entirely: 'There is no such thing as *general* intelligence ... Our environment, which determines how our intelligence manifests itself, puts a hard limit on what we can do with our brains.' [60]

15. At least in retrospect, AV pioneer Ernst Dickmanns explicitly categorized his development of self-driving vehicles as intelligence research. He defined intelligence as the ability to gain insights into processes in the real world, based on a background understanding and the capacity to transfer knowledge. This knowledge of the physical world, encompassing the vehicle's body itself, enables the system to handle a wide range of situations that cannot be conclusively described in advance. The visionary concluded: 'In front of this background, it seems reasonable to equip complex technical systems with a similarly advanced sensor suite as humans have, with an interpretation background on the latest state of development in the natural sciences and engineering ... With this Information and with digital maps of the continents, technical autonomous Systems will have global navigation capabilities far exceeding those of biological Systems.' [90]

16. In fact, Lovelace was the first to identify creativity as a core criterion for machine intelligence, albeit in

a sceptical way in Note G, 'The Analytical Engine has no pretensions whatever to originate anything. It can do whatever we know how to order it to perform. It can follow analysis, but it has no power to anticipate any analytical relations or truths. Its province is to assist us in making available what we are already acquainted with.' [16] Indeed, Alan Turing referenced her objection—that machines only do what they are programmed to do—in his seminal 1950 paper: 'The Analytical Engine has no pretensions to originate anything. It can do whatever we know how to order it to perform.' [350] This inspired Selmer Bringsjord, Paul Bello and David Ferrucci to introduce a *Lovelace Test*, specifically designed to assess originality. [46] Interestingly, Lovelace demonstrated foresight in this regard as well, as creativity is now the legal criterion for copyrightable authorship in AI-generated artworks. [227]

17. Turochamp, developed by Turing and Champernowne, is considered the earliest computer game in development. However, it was never completed due to its complexity surpassing the capabilities of contemporary computers. In 1952, Turing manually executed the program to play against Alick Glennie. Despite this, Turing died in 1954 without ever running Turochamp on a real computer.

18. In contrast, Jürgen Schmidhuber sees much deeper roots to AI: foregrounding the chain rule for backward credit assignment from 1676; linear regression, a form of shallow learning, to around 1880 and dates the first recurrent neural network architecture to between 1920–1925. [308]

19. Kybernetes, meaning steersman, governor, pilot, or rudder, and having the same root as *government*.

20. For a vivid description of Walter Pitts' unusual career and his contribution to the invention of artificial neural networks, see Anil Ananthaswamy (2024). [9]

21. More formally stated, McCulloch and Pitts proved that a class of logical expressions (including predicates that take predicates as arguments), which they called temporal propositional expressions, (TPEs), have a single free variable and are recursively defined such that (a) any predicate of one argument is a TPE; (b) the logical conjunction (and), disjunction (or) and negated conjunction (and not) of any two TPEs with the same variable is a TPE; (c) a TPE in which the value of the time variable is substituted by its predecessor (time delay) is a TPE, and nothing else is a TPE, then any TPE is realizable by a non-recurrent network. That is, there will always be a network of MCP cells whose synapses or connections do not form cycles, and which can compute any given TPE. Indeed, some years later, first Stephen Kleene [185]. Subsequently, Marvin Minsky [245] proved that rational-weighted recurrent neural networks with Boolean activation functions are computationally equivalent to classical finite state automata.

22. In a simple McCulloch–Pitts threshold model, adaptability comes from representing each synaptic junction by a variable—usually rational—valued weight W_i, indicating the degree to which the neuron should react to the $_ith$ particular input (see Figure 2.1). By convention, positive weights represent excitatory synapses and negative inhibitory synapses, and a variable T represents the neuron firing threshold. Typically, in modern usage, T is clamped to zero, and a threshold is implemented using a variable *bias* weight, b.[28] A neuron firing is usually represented by the value $+1$ and not firing by 0.[29]

    Activity at the i_{th} input to an n input neuron is represented by the symbol X_i, and the effect of the i_{th} synapse by a weight W_i, hence the net effect of the i_{th} input on the i_{th} synapse on the MCP cell is thus $X_i \times W_i$.

    Thus, the MCP cell is denoted as firing if:

$$\sum_{i}^{n} X_i \times W_i + b \quad \geq \quad 0 \tag{2.1}$$

23. Example activation functions include: *linear summation*; the *Heaviside* (unit step or threshold) function, usually denoted by H, which is a discontinuous function whose value is 0 for a negative argument and $+1$ for positive argument; the *rectified linear unit* (ReLU), and the widely deployed *sigmoid* function (which offers a continuous, differentiable approximation to the Heaviside function).

24. According to Cassirer, the symbol does not have a physical existence. In Cassirer's philosophy of symbolic forms, symbols are a product of human consciousness and language, and they are distinct from physical objects or events. Cassirer argued that symbols are the means by which human beings understand and interpret the world around them and that they are essential to human thought and culture. Therefore,

symbols have a different mode of existence than physical objects and cannot be reduced to their physical components or properties. Thirty-six years later, John Searle reached a similar conclusion via the Chinese Room Argument (CRA). [314] This has critical implications for AI and any computational theory of mind.

25. A *heuristic* is simply a function that estimates the cost or distance from a current state to the goal state.

26. Sensu stricto, hybrid, *so-called neuro-symbolic* systems, are also investigated.

27. Mathematical objects used to represent multi-dimensional real-valued data that generalize scalars, vectors, and matrices to higher dimensions.

28. A bias weight is simply a weight connected to an input clamped to $+1$.

29. In psychology, '*the fundamental relations are those of two-valued logic*', and McCulloch and Pitts recognized neuronal firing as equivalent to representing a proposition as $TRUE$ or $FALSE$. [236]

Section II

Symbols and Rules

In *Driving Intelligence*, we identify the *first wave of vehicular autonomy*, leading up to the initial DARPA Grand Challenge for autonomous vehicles in 2004, as being predominantly defined by AI systems built using symbols and rules rather than machine learning. This preference was shaped by the technological, computational, and theoretical constraints of the time, as well as by prevailing paradigms in artificial intelligence and robotics.

At this time, machine learning was neither as advanced nor as accessible as it is today. Many of the modern ML techniques, particularly deep learning approaches, were either still in their infancy or had not yet been developed. Existing ML algorithms proved unable to cope with the complexity and variability inherent in real-world driving conditions. They typically required carefully handcrafted features, making them brittle and less adaptable to new environments. Furthermore, the computational power necessary to train and deploy ML systems, particularly for real-time applications, was not widely available. Hardware such as GPUs and specialized processors for ML was either unavailable or prohibitively expensive, limiting the feasibility of using data-driven approaches for autonomous driving.

Data availability was another significant constraint. Machine learning relies on large data sets for training and testing; however, prior to the DARPA Challenges, there were few high-quality data sets available for tasks such as road navigation or obstacle avoidance. Building the large and diverse data sets required to train ML models effectively was not practical at the time, and this lack of data made it challenging for ML approaches to gain traction in autonomous vehicle research.

During this period, symbolic AI and rule-based systems dominated the field. These approaches were rooted in explicitly encoding knowledge as logical rules and symbols, which aligned with the engineering practices of the era. Autonomous systems were designed with deterministic control and logic-based frameworks, where developers could manually define specific behaviours, such as detecting obstacles or following a road. Rule-based systems were seen as more interpretable, predictable, and verifiable compared to the statistical methods underlying machine learning. This explicitness made them easier to debug and validate, particularly for safety-critical tasks like driving, where failure can have the most severe consequences.

Real-time decision-making also poses a significant challenge. Autonomous vehicles require control systems that can operate efficiently within the limited computational resources of the time. Rule-based systems were computationally lightweight and could meet these constraints, whereas early ML models often demanded significant processing power and memory, making them unsuitable for real-time deployment.

Another critical factor was the general scepticism and lack of trust in learning-based systems for safety-critical applications. Rule-based systems provided a sense of control and predictability as well as mathematical provability, [319], because their behaviours could be explicitly programmed and exhaustively tested for specific scenarios. In contrast, ML systems, which learn from data and exhibit unexpected behaviours, are challenging to interpret and verify, due to their black box nature. This lack of transparency made engineers and researchers hesitant to rely on ML for tasks like autonomous navigation.

The immaturity of perception and sensor technology further influenced this dynamic. Sensors available for autonomous vehicles at the time—such as cameras, lidar, and GPS—were less sophisticated than those used today. The limited accuracy and richness of data provided by these sensors constrained the potential of ML models, making simpler processing pipelines based on geometric reasoning and symbolic rules more practical and reliable.

However, as we unfold, the first DARPA Grand Challenge already highlighted the limitations of rule-based systems as the vehicles struggled to navigate the open, unstructured environments of the desert test bed. This demonstrated the advantages of more adaptive and data-driven approaches. Over the following years, advances in ML algorithms, the availability of large data sets, improvements in sensor technology, and the development of high-performance computing hardware paved the way for a transition toward machine learning-based methods. Today, ML systems are the leading technology in autonomous vehicles. They initially gained prominence in the vision component of the traditional sense-plan-act architecture. However, with the emergence of end-to-end learning architectures, handcrafted algorithms in the planning and actuation modules are becoming increasingly obsolete—at least in Level–2 systems.[1] During the early days of the first DARPA challenge, learning from vast amounts of data, adapting to diverse scenarios, and achieving levels of performance were unimaginable.

3 First Steps in Computer Vision

Vision has long been integral to the development of autonomous vehicles, as exemplified in the first autonomous intelligent car built by Sadayuki Tsugawa and his colleagues in 1977 at the University of Tsukuba's Mechanical Engineering Lab in Japan; a vehicle which used two video cameras and analogue signal processing to detect street markings, and which could autonomously drive on roads at speeds of up to 30 km/h (see Section 1.4). [347] If machines could *see* and interpret the world as effectively as humans, the challenge of autonomous driving would be much easier. This is the ultimate goal of computer vision, a branch of artificial intelligence dedicated to enabling machines to analyse and interpret visual data. [356]

For computers, visual data in the form of digitized images are defined by two-dimensional arrays of numbers, encoding atomistic image elements called *pixels*, where each pixel represents a point in the image and has a scalar value. In greyscale images, the pixel values range from 0 (the darkest) to 255 (the brightest). For colour images, the red, green, and blue (RGB) values of each pixel are represented as a list of three values between 0 and 255 (See Figure 3.1).

One of the earliest theories of vision, the template matching model, was developed to explain how patterns can be recognized. This model suggests that external stimuli are compared pixel by pixel to internal mental representations, or *templates*, and recognition occurs when there is sufficient overlap between the stimulus and its corresponding template. However, it was immediately apparent that—*as a theory of vision*—template matching has fundamental flaws. Not least, it fails to account for image constancy—the ability to recognize stimuli despite variations in environmental conditions.

For example, a bold **Q** and a regular *Q* have distinct pixel configurations. However, it is implausible that humans store a separate template for every possible variant of a pattern. In addition, changes in lighting and background can alter pixel values in images of the same object. This issue is particularly problematic for patterns with minor pixel differences, such as the numerals $0, 3$, and 8, or letters like O and Q, where such variations can lead to misclassification. Furthermore, even very slight changes in orientation and depth can significantly alter the two-dimensional projection of a three-dimensional object as captured by a camera.

Historically, this problem of image constancy has been addressed by an intermediate image processing step called feature extraction, which entails using algorithms that endeavour to detect elements of an image in a way that is invariant to environmental changes before the object is recognized as a whole.[2] One of the first and best-known feature detection models was Oliver Selfridge's Pandemonium.

DOI: 10.1201/9781003561460-5

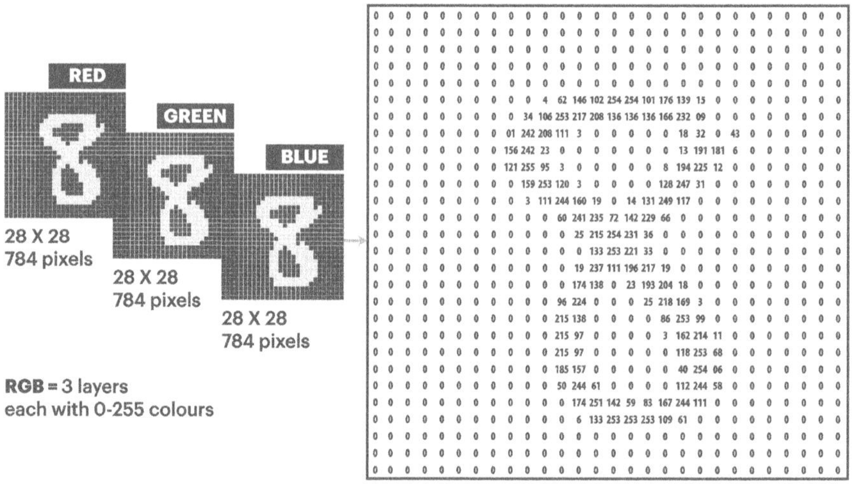

Figure 3.1 Images as pixels. (Image adapted from Pritam Basu, 2018 [23].)

3.1 THE FIRST COMPUTER VISION SYSTEMS

Since its inception, computer vision has been a central focus of AI, aiming to automate the classification of elements within visual scenes. Two closely related early contributions to this field were Oliver Selfridge's 1955 paper, 'Pattern recognition and modern computers', [316] and G.P. Dinneen's 1955 paper, 'Programming pattern recognition'. [92] Both were presented at the '1955 Western Joint Computer Conference' and addressed the emerging and cutting-edge area of pattern recognition in mid-20th-century computer science.

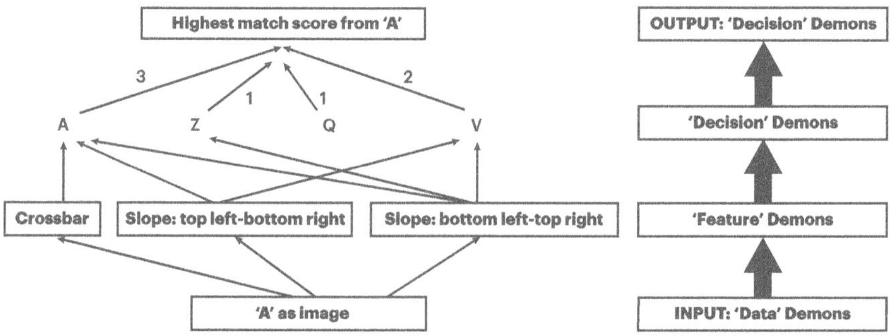

Figure 3.2 Pandemonium identifying the character '*A*'.

Selfridge subsequently refined his ideas into Pandemonium: a paradigm for learning, a conceptual framework for pattern recognition inspired by cognitive processes, which he presented at the 1958 Teddington 'Mechanisation of Thought Processes

Symposium'. [315] In Selfridge's Pandemonium, pattern recognition is envisioned as a *hierarchical* process performed by four metaphorical *demons*, each specializing in a specific task. The final interpretation of a scene arises from the collective activity of numerous simple processors, notionally operating in parallel.

'Image daemons' handle raw sensory input, forming the initial stage of data acquisition. Feature daemons then identify core image elements such as lines, angles, or curves, responding when their assigned feature is present. Cognitive daemons represent potential patterns or objects and become increasingly active as more of their associated features are identified. Finally, the decision daemon selects the loudest (most active) cognitive daemon, determining the recognized pattern.

Pandemonium can be viewed as an early form of a machine learning system, as it adjusts the strength of responses between daemons based on probabilities derived from overall pattern-matching outcomes, effectively a simplified form of reinforcement learning. Bolstering this analogy, the term *demon* is a slant rhyme with *neuron*, further suggesting its role in a connectionist-like process.

Through repeated exposure and reinforcement of correct identifications, the model refines these weights, suppressing incorrect responses. For instance, when learning to recognize the letter '*A*', the model might initially confuse it with '*V*' due to their similar diagonal features. However, with continued exposure, the weights adjust, and the '*V*' response is inhibited whenever distinctive features of '*A*', such as a crossbar, are identified.

Though innovative in its capacity to generate and discard characteristics based on utility, the Pandemonium model had limitations, particularly its tendency to overemphasize a single feature. Despite these shortcomings and the model's simplicity compared to modern approaches, Pandemonium was a groundbreaking demonstration of how emergent behaviour arising from the interaction of multiple distributed and parallel processes could support pattern recognition.[3] Its subsequent influence extended to fields such as perception, computer vision, and machine learning approaches to AI.

Larry Roberts significantly advanced the emerging field of computer vision in the 1960s with his research in scene analysis. His work focused on understanding two-dimensional images of three-dimensional scenes composed of various assemblies of polyhedra. [281] Roberts' program identified objects, calculated their dimensions, and determined their absolute and relative positions, even when partially obstructed. It could reconstruct views of objects from different perspectives but required detailed mathematical descriptions of all objects in its Blocks World to function.

In its simplest form, Roberts' program categorized objects into three classes: *cuboids*, *wedges*, and *hexagonal prisms*, which could combine to form *compound shapes*. The program used geometric interrelations between surfaces, edges, and corners to define models and evaluated the match between models and image data based on a numeric similarity score. The system combined bottom-up and top-down processing, using cues from image subsections to suggest models and confirming these models through further cue analysis. However, despite its geometric precision, the program was computationally inefficient due to its use of exhaustive search algorithms, and it struggled to process anything other than simple images.

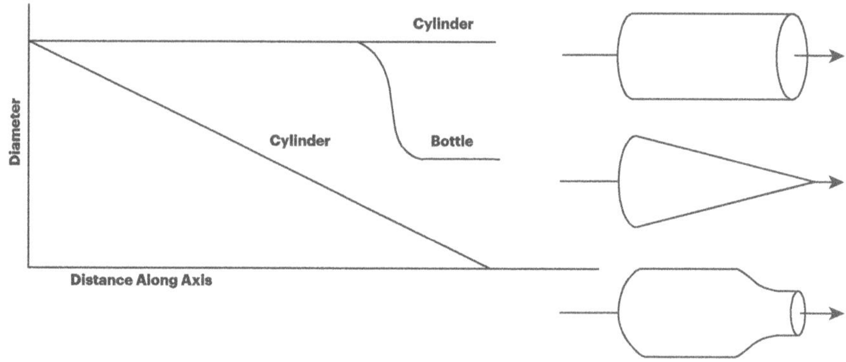

Figure 3.3 Binford's generalized cylinder model for describing shape and form. (Image adapted from Bishop 1989 [32].)

Adolfo Guzmán's work in the late 1960s advanced scene analysis with the development of SEE, a system that isolated 3D bodies from 2D images without relying on specific model data. [124] SEE used a bottom-up approach to extract object regions and vertices, linking these regions to form nuclei. Although less precise than Roberts' program, it was more general, as it did not depend on predefined models.

Subsequent contributions by Huffman [157] and Clowes [68] enhanced scene analysis through their work on line drawings. They focused on interpreting junctions and edges to infer the underlying 3D structure and introduced labelling strategies to address nonsensical or impossible objects. However, despite these improvements, significant limitations persisted, particularly the inability to handle partially occluded objects.

However, by the 1970s, Waltz had demonstrated that exhaustive searches of local arrangements of surfaces, edges, and shadows could yield practical image interpretation algorithms, marking a culmination of this phase of research. [366] Nonetheless, these programs were constrained by their restricted problem domain and lacked an understanding of physical interactions in the scene, e.g. the support that a cube provides to a cone placed on top of it. Despite these limitations, as Marr later observed, this research offered foundational insights into how to proceed once lines/edges had been extracted from a scene.

Recognizing the need for more complex object representations, in 1971, Binford introduced the concept of generalized cylinders (see Figure 3.3). [29] This model accommodated variations in shape and axis, expanding the representational scope of object recognition systems and allowing for more realistic descriptions of shapes beyond the rigid polygons of the early Blocks Worlds.

In parallel with these theoretical advancements, empirical research, such as Rosenfeld and Thurston's work on edge detection, explored practical methods for processing real-world images. A particular challenge was distinguishing surface reflectance changes from lighting variations. Here, one approach, derived from Land and McCann's retinex theory of colour vision, proposed filtering the gradual changes in illumination to isolate the sharp changes caused by reflectance. [194] Marr and Poggio later built on this insight, emphasizing the need to understand information processing tasks in perception independently of the mechanisms implementing them. [225] This idea fundamentally informed Marr's subsequent, highly influential theories on vision. [223]

In summary, the earliest *symbolic*, or rule-based, approaches to computer vision involved representing visual elements as discrete symbols and relationships, which were then manipulated using formal logic and reasoning. Unlike machine learning methods that rely on extensive training data sets, symbolic vision systems utilize predefined rules about object properties, spatial relationships, and hierarchical classifications to interpret visual information. This methodology enabled AI systems to perform visual reasoning tasks with limited data, offering interpretability and simple reasoning capabilities in early machine vision systems.

3.2 DAVID MARR'S APPROACH TO COMPUTER VISION

Perhaps the single most influential figure in early computer vision was David Marr from the Massachusetts Institute of Technology (MIT). His approach to computer vision was fundamentally symbolic and grounded in computational theory rather than, say, machine learning. [223] However, his foundational insights are used across schools, especially in convolutional neural networks that underlie many autonomous vehicles. Marr proposed a hierarchical framework for understanding vision, which consisted of three levels of abstraction: the raw, *primal sketch*, capturing intensity changes such as edges and contours; *the 2.5D sketch*, representing surfaces and depth from the viewer's perspective; and *the 3D model*, encoding objects in a viewer-invariant manner. These levels relied on explicit *symbolic representations* to describe visual information.

Marr's approach emphasized the *algorithmic processes* that operate on such *representations* to extract meaningful *features*. For example, the Marr-Hildreth edge detection algorithm employed the Laplacian of Gaussian filter to highlight edges mathematically. [224] Thus, Marr treated the vision system as a modular series of symbolic processes, each designed to solve a specific task, focusing on explaining how and why the visual system works.

Unlike machine learning approaches, which learn representations directly from data, Marr's work relied on manually defined algorithms and logic, reflecting the technological constraints of his time. While his methods predated modern data-driven approaches, his ideas on hierarchical processing and feature extraction had a lasting influence on the field. In broad summary, the strategy Marr defined can be outlined as follows [223]:

1. Edge-filter the image.
    - Edges are defined as sudden intensity changes which give rise to a peak or trough in the first derivative and a zero crossing in the second.
    - The filter that Marr argues is the most satisfactory engineering approximation to the ideal filter is the DOG (Difference of Gaussians). [223] [224]

2. Construct the raw primal sketch.
   - This is an image constructed using edge, bar, blob, termination, and discontinuity primitives.
3. Construct the full primal sketch.
   - This image is built up from descriptive primitives in a recursive manner to capture spatial structure at each scale, using the same primitives as the raw, primal sketch. For example, in an image of a cat, the lowest level of structure may be individual hairs, while the highest level is the overall shape of the cat.
4. From images to surfaces.
   - This entails the use of various processing modules, such as stereopsis, directional selectivity and apparent motion, to gain local depth and other constraint information.
5. Construct the 2.5D sketch.
   - The 2.5D sketch can be derived by combining information from the various processing modules. It makes explicit information on the orientation and rough depth of the visible surfaces and the contours of discontinuities of these quantities in a viewer-centred coordinate frame (See Figure 3.4).

The 2.5D sketch retains the viewer's perspective and encodes depth and surface orientation relative to the observer. This intermediate representation is crucial for perceiving spatial relationships and surfaces; however, it is not sufficient for object recognition, as it remains viewpoint-dependent. Prior to object recognition, the 2.5D

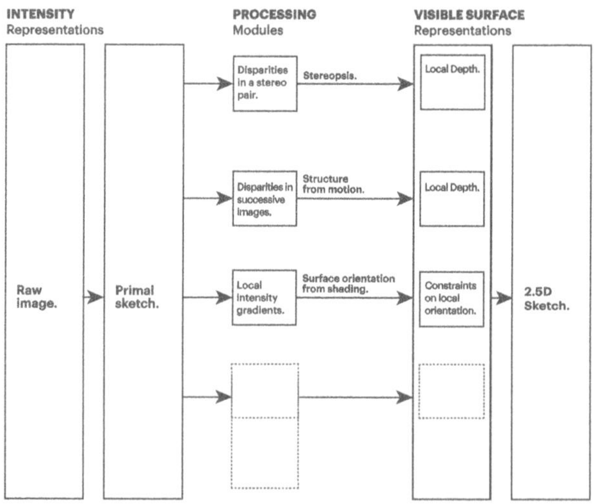

Figure 3.4 The formation of Marr's 2.5D sketch. (Image adapted from Bishop 1989 [32].)

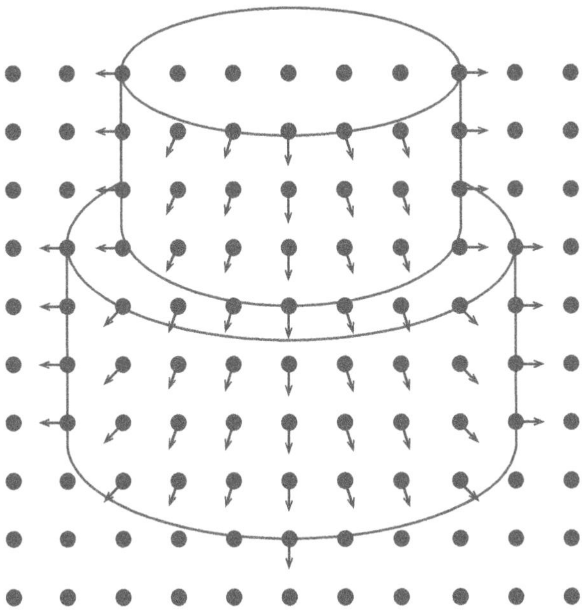

Figure 3.5 An example 2.5D sketch—where depth and orientation are explicit. (Image adapted from Bishop 1989 [32].)

sketch (See Figure 3.5) must be transformed into an *object-centred* representation that depends minimally on the viewpoint.

In Marr's theory, object recognition occurs at the level of the 3D model, where objects are represented in a canonical form independent of the viewer's perspective. This 3D model is constructed by integrating information from the 2.5D sketch and comparing it to stored structural descriptions of known objects. These descriptions are not simple *templates* but are complex hierarchical models detailing an object's parts and their spatial relationships, enabling flexible recognition across varying viewpoints.

Building on Marr's insights, early symbolic approaches to 3D vehicle recognition, e.g. by Baker and Sullivan at the University of Reading Computer Vision Group, [19]) evolved by utilizing the following principles:

1. Edge-filter the raw image.
2. Extract key 2D features.
3. Constrain these features, e.g. using an expert system.
4. Construction of an object-centred representation (*view-sphere*).
5. Back projection of the view-sphere and evaluation against raw image data.

Rule-based symbolic approaches, such as these, were those predominantly used in the development of early autonomous vehicles leading up to the first DARPA Grand Challenge in 2004. Fundamentally, these methods relied on identifying predefined features, such as edges, shapes, and colours, to detect and classify objects within the environment.

3.3 3D VISION: STRUCTURE FROM MOTION

As computer vision algorithms improved, researchers began to explore new ways to expand their capabilities in the realm of three-dimensional images. One of the most critical developments in the field was structure-from-motion (SfM) techniques.

Structure-from-motion enables the 3D structure of a scene to be reconstructed from a series of 2D images taken from different viewpoints. It involves estimating the motion and depth of points in a scene from their projections in multiple photos taken from various perspectives. By analyzing the relative motion of these points across different images, a computer can create a 3D model of the scene. This technique is widely adopted in applications such as autonomous driving and robotics.

The development of *Structure-from-motion* (SfM) has been influenced by the works of many researchers in computer vision; some of the key pioneers who contributed were:

Longuet-Higgins: In 1981, he introduced the concept of 'structure from motion,' which involves recovering the 3D structure of a scene from a sequence of images taken from different viewpoints. [214]

Tomasi and Kanade: In 1992, they introduced an approach to SfM based on factorisation that decomposes a sequence of images into the intrinsic parameters of the camera and the 3D structure of the scene. [344]

Torr and Zisserman: In 2000, in a paper entitled 'Feature-based methods for estimating structure from motion' an algorithm called 'Affine Structure from Motion' estimates the 3D structure of a scene from a set of 2D images, even when the photos have been taken from different viewpoints and under various lighting conditions, based on the assumption that the affine structure of the scene remains constant across the images.[4] [345]

Stewnius, Engels and Nistr: In 2006, they proposed a novel version of the '5-point algorithm' that can recover the 3D structure of a scene from a sequence of images taken from a single moving camera. [334]

Snavely, Seitz and Szeliski: In 2006, they introduced an SfM system that can create 3D models from an extensive collection of unordered photographs available on the internet. [330]

Computer vision, which has its roots in the very early days of computer science and artificial intelligence, has come a very long way from the analysis of satellite images developed in the 1950s at the Jet Propulsion Laboratory, Caltech, from two-dimensional image segmentation to object detection and three-dimensional image understanding; the ability of machines to *see* and *interpret* the world around them is on its way to becoming reality.

Notes

1. Level–3 and Level–4 systems with end-to-end approaches are currently not allowed on public roads, given the fundamental opacity of deep neural networks.

2. One method to mitigate changes due to changing lighting intensity, suggested by Paul Minchinton et al., is via the use of *Minchinton cells*, as these process the *relative* intensity of pairs of pixels in an image. [36]

3. The debate around emergent abilities in LLMs questions whether these capabilities truly emerge at scale or are artefacts of measurement choice. Schaeffer et al. argue the latter, demonstrating that non-linear metrics create apparent emergence while linear metrics show smooth, predictable performance changes. [298] Lu et al. suspect in-context learning to be behind the supposed convergent abilities of large neural networks: 'Our findings suggest that purported emergent abilities are not truly emergent, but result from a combination of in-context learning, model memory, and linguistic knowledge.' [218] We concur that there is no apparent reason to assume an inexplicable *emergence* to solve open technical questions by themselves miraculously. Occam's razor demands us to look for the most straightforward answers when explaining observations.

4. Professor Torr has also worked on developing algorithms for robustly estimating the structure of a scene in the presence of outliers and for integrating structure-from-motion with other computer vision tasks such as object recognition and tracking. His work has been extremely widely cited and has helped to advance the field of structure-from-motion.

4 The Tortoise and the Hare

The development of self-driving cars and autonomous wheelchairs (AWs) represents two streams of innovation in AV research that started from different points but have increasingly converged. The shared goal—as they are both vehicles—is to achieve full autonomy, enabling independent operation without human intervention, albeit with varying levels of complexity and safety requirements.

In the early days, both AWs and AVs deployed AI systems defined by a corpus of *rules*, operating on *symbolic* knowledge structures, using navigation information loaded from *maps* and data *sensed* from the local environment. They subsequently progressed through comparable developmental stages, with the first systems focused on basic automation, such as obstacle avoidance for wheelchairs and automatic braking in cars. Over time, both fields advanced to incorporate more sophisticated features, such as adaptive cruise control in AVs and sophisticated path-planning systems in AWs.

Thus, while their technologies serve different primary purposes—transportation for the general public versus mobility assistance for individuals with disabilities—AVs and AWs share a common foundation in artificial intelligence, computer vision, machine learning, robotics, and sensor technologies. This overlap has led to a synergistic relationship, where advancements in one field often benefit the other, and one of the most important areas of crossover technology has been *sensors*.

4.1 SENSOR TECHNOLOGIES FOR AUTONOMOUS NAVIGATION AND CONTROL

Many sensors, such as odometry, infrared, ultrasonic, radar, lidar, and cameras, initially deployed in self-driving cars, have also been applicable to autonomous wheelchair systems.[1] However, in both applications, each technology has its own merits and demerits. As a result, data from one sensor is often combined with data obtained using complementary technology(s) in a process called *sensor fusion* to provide a better description of the environment.

Odometry: Odometry sensors in autonomous vehicles measure movement, speed, and direction by tracking changes in position, typically through wheel encoders or inertial measurement units (IMUs). These sensors are crucial for real-time position tracking, enabling the vehicle to navigate by estimating its location relative to a starting point. They are instrumental in situations where GPS signals are weak or unavailable, such as tunnels or urban areas. However, odometry sensors can accumulate errors over time due to factors such as wheel slip or rough terrain, resulting in inaccuracies in position estimates.

Accelerometers: An accelerometer is a sensor that measures acceleration, which is the rate of change of velocity over time. These devices can detect acceleration in

DOI: 10.1201/9781003561460-6

one, two, or three dimensions, depending on their design, and provide data on both static forces, such as gravity, and dynamic forces, such as movement or vibration. Accelerometers work by measuring the displacement of a small internal mass in response to external forces. This displacement is detected using technologies such as piezoelectric materials, capacitive sensing, or microelectromechanical systems (MEMS). The sensor converts the displacement into an electrical signal, which is processed to determine the magnitude and direction of acceleration.

Gyroscopes: A gyroscope sensor measures angular velocity, capturing the rate of rotation around one or more axes. Using technologies such as mechanical, optical, or microelectromechanical systems, gyroscopes detect changes in orientation and provide precise rotational data. They work by exploiting the principles of angular momentum, where the rotation of an internal component remains stable relative to external forces, allowing the sensor to measure shifts in orientation.

In autonomous vehicles, gyroscope sensors are integral to navigation and stability systems. They provide real-time data on the vehicle's orientation, enabling precise control during manoeuvres and assisting in maintaining balance. Combined with accelerometers, they form an inertial measurement unit (IMU), which tracks changes in position and orientation without relying on external signals. This is particularly critical in GPS-denied environments, such as tunnels or areas with poor satellite coverage.

Ultrasonics: Ultrasonic sensors, which emit high-frequency sound waves and measure the time for the echo to return, are widely used in autonomous vehicles for close-range object detection and obstacle avoidance. Commonly employed for tasks like parking assistance, low-speed obstacle detection, and blind spot monitoring, they offer high accuracy within a few meters and can be easily integrated into vehicle bumpers or mirrors. Their reliability in adverse weather and poor lighting conditions makes them especially useful in low-speed scenarios. However, their short detection range and sensitivity to environmental noise or soft objects limit their effectiveness at speed.

IR: **I**nfra**r**ed sensors in autonomous vehicles are used to detect heat emitted by objects, enabling them to function effectively in low-visibility conditions like darkness, fog, or rain. They are most commonly used to detect pedestrians, animals, and other obstacles by identifying their thermal signatures.

Radar: **Ra**dio **d**etection **a**nd **r**anging sensors in autonomous vehicles use radio waves to detect objects and measure their distance, speed, and direction, making them essential for medium to long-range detection in various driving conditions. They are particularly effective at high speeds, which are crucial for tasks like motorway driving and adaptive cruise control (ACC), where they help maintain a safe distance by adjusting speed in real-time. Radar also plays a key role in collision avoidance systems by detecting obstacles and triggering emergency braking or steering, and it assists in lane changes by monitoring nearby vehicles. Radar sensors provide accurate measurements of speed and distance, perform well in darkness and all weather conditions, and are especially useful for high-speed applications. However, they have limitations, such as lower resolution compared to lidar or cameras, and can be affected by interference from other radar signals or environmental factors.

lidar: **Li**ght **d**etection **a**nd **r**anging sensors use laser pulses to measure distances by calculating the time it takes for light to reflect off objects, providing real-time, high-resolution 3D maps essential for navigation and decision-making in autonomous vehicles. Thus, lidar sensors generate a detailed point cloud of data consisting of distance vectors that measure the distance to the first reflective surface, allowing for precise environmental scanning and accurate path planning. Additionally, lidar performs well in low-light conditions, such as at night or in tunnels, making it essential for understanding the vehicle's surroundings. Hence, lidar technology facilitates the detection of obstacles, such as pedestrians, cyclists, and other vehicles, while also identifying lanes, road edges, and traffic signs.

A limitation of early scanned lidar devices is that they only collected point data in a single plane. Although two-axis mechanically scanned lidars with an acceptable range have now been available for several decades, the requirement to scan on two axes limits their scan rate. This, however, is rarely a limitation in automated wheelchair applications.

Nonetheless, lidar has other disadvantages, including its high cost and sensitivity to visible airborne particles, like dust, snow, heavy rain, and dense fog, which can scatter or block laser pulses. Additionally, it has a shorter detection range than radar, limiting its effectiveness at high speeds.

Vision: *RGB cameras* can capture detailed colour images, facilitating object detection and sign identification. However, their performance can be compromised in low-light conditions. Infrared cameras, also known as thermal cameras, can detect heat signatures, allowing for the identification of people and animals even in complete darkness. *Event cameras* can respond to changes in a visual scene rather than capturing entire frames, allowing for high-speed motion detection with low latency. This capability is advantageous for tracking fast-moving objects and reducing data processing requirements. *Range-gated cameras* can control the time window during which light is detected, capturing images at specific distances; a feature is handy in poor light conditions. *Polarisation cameras* can detect the orientation of light waves, providing information about surface properties and enhancing visibility in challenging situations where there is glare or reflections. The primary disadvantage of cameras is that they do not perform physical measurements, unlike radar or lidar. If objects are not classified by a vision algorithm that evaluates the data, objects will be treated as non-existent. Additionally, cameras only capture 2D images. For three-dimensional structures, two or more lenses often work together, which can lead to susceptibility during calibration and positional shifts due to vibrations or general deterioration over time. Cameras are also sensitive to soiling and other impairments to visual quality, such as glare, optical reflections, or weather effects.

GPS: GPS sensors are devices that use signals from the global positioning system (GPS), a satellite network, to determine precise locations on earth. The GPS operates through a constellation of at least 24 satellites that broadcast signals containing their positions and the exact time of transmission. GPS sensors receive these signals and calculate their position using a process called triangulation. By measuring the time delay between signal transmission and reception, the sensor determines its distance from each satellite and, using data from at least four satellites, computes its three-dimensional position (latitude, longitude, and altitude) as well as time. GPS sensors provide accurate location data, typically within 3–10 metres for civilian use. However, advanced methods like differential GPS and real-time kinematic positioning can improve accuracy at centimetre levels. However, the main disadvantage of GPS is that it requires a clear line of sight to the connecting satellites at all times. The accuracy of the sensor depends on the quality of the received signal, which does not penetrate solid walls or structures. Even a dense forest or high-rise buildings at the side of the road pose a problem. GPS sensors are designed for efficiency and are often integrated with other systems, such as accelerometers or gyroscopes, to enhance performance in challenging environments like cluttered urban areas.

Similarly, the sophisticated rule-based, symbolic AI algorithms used for computer vision, path-planning algorithms and obstacle avoidance, initially developed for AVs, have also been adapted to help AWs navigate complex indoor and outdoor environments.

Conversely, the specific needs and constraints of wheelchair users have driven unique innovations in AW design. These include specialized human-computer interfaces for users with limited mobility, more compact sensor arrays for navigating tight indoor spaces, and AI systems tailored to understand and anticipate the needs of individuals with disabilities. In particular, and in stark contrast to AV localization systems, a wheelchair positioning system typically has to operate in a highly cluttered and dynamic indoor environment where technologies like GPS can, at best, only offer minimal assistance.

4.2 THE EVOLUTION OF AUTONOMOUS WHEELCHAIRS

Throughout the 1990s and into the early 2000s, during the period of AV development Ernst Dickmanns described as a virtual AI winter caused by the lack of computing power available to underpin AV systems (see page 13), remarkable progress was made towards building viable AWs. Primarily, this was because AWs operate an order of magnitude slower than AVs and hence require significantly less computing power to react to their environment in real-time.

In 1986, at the height of this virtual AI winter and while computer vision remained in relative infancy, work on AWs began to gain broader attention with the construction of a vision-aided system for wheelchair navigation support. The smart wheelchair programme at the CALL centre, University of Edinburgh (UK), developed a prototype with collision avoidance features. [219] This prototype was based on integrated proximity sensors and odometry, facilitating adaptation to various use cases. [72]

Considerable interest also arose in so-called *assistive robotics*, primarily focused on semi-autonomous wheelchair navigation systems. In 1990, Borenstein et al. [42]

presented the development of a wheelchair prototype based on the *vector field histogram* method, which was subsequently extended in the implementation of the obstacle avoidance and risk management for the SENARIO wheelchair (described in the following section).

Concomitant progress was being made in the areas of rehabilitation robotics and advanced wheelchairs throughout Europe. [76] Thus, in 1995, the OMNI project developed a novel *omnidirectional* wheelchair that featured navigational features integrating both infrared and ultrasonic sensors. [154] Indeed, the increased interest and investment in the field at this time led to two notable international competitions: the wheelchair competition held at the 'Association for the Advancement of Artificial Intelligence' conference in 1996, and the robot wheelchair competition held during IJCAI-95. [239]

4.3 THE SENARIO AUTONOMOUS WHEELCHAIR

In the mid-1990s, the European Union directed funding towards research in the areas of home robotics and autonomous vehicles for healthcare through the *Technology Initiative for the Disabled and Elderly* (TIDE) programme. [100] One influential project funded by this initiative was the *Sensor-Aided Intelligent Navigation System for Powered Wheelchairs*, commonly known as SENARIO. [177] The objective of this pan-European project was to develop a sensor-aided intelligent navigation system to provide high-level navigational assistance to users of powered wheelchairs, supporting both semi-autonomous and fully autonomous navigation.[2]

The challenges posed by the cluttered and dynamic environments of hospitals and homes, where the wheelchair had to operate autonomously, required a collaborative effort from several research groups across Europe, each developing specific components of the core system architecture.

SENARIO implemented its autonomous vehicle behaviour through four core modules: *environment perception* (sensing, positioning, and the user interface), *risk management*, *actuation*, and *task planning*. The task planning module, which oversaw the execution of all other processes, was the highest in the control hierarchy, thus centralizing the vehicle's control (although direct communication between lower-level processes was available in emergencies).

Because the SENARIO wheelchair was designed for use in indoor environments, it was unsuitable for GPS technology. As a result, a new method had to be designed to position and orient the vehicle robustly from scratch. At the time, neither computer vision algorithms nor hardware were advanced enough to perform real-time localization in an unconstrained, noisy, real-world environment. Therefore, the SENARIO localization team opted to use a map of the locale, lidar sensors, and stochastic diffusion search, [30] a bespoke swarm intelligence algorithm (see page 75), to interpret lidar sensor range data and position and orient the vehicle within a room (see Figure 4.1). [24]

The SENARIO project employed a localization subsystem based on two SICK-180 infrared diode range scanners, which generated a 720-element range vector as output. The goal of the subsystem was to determine the position and orientation

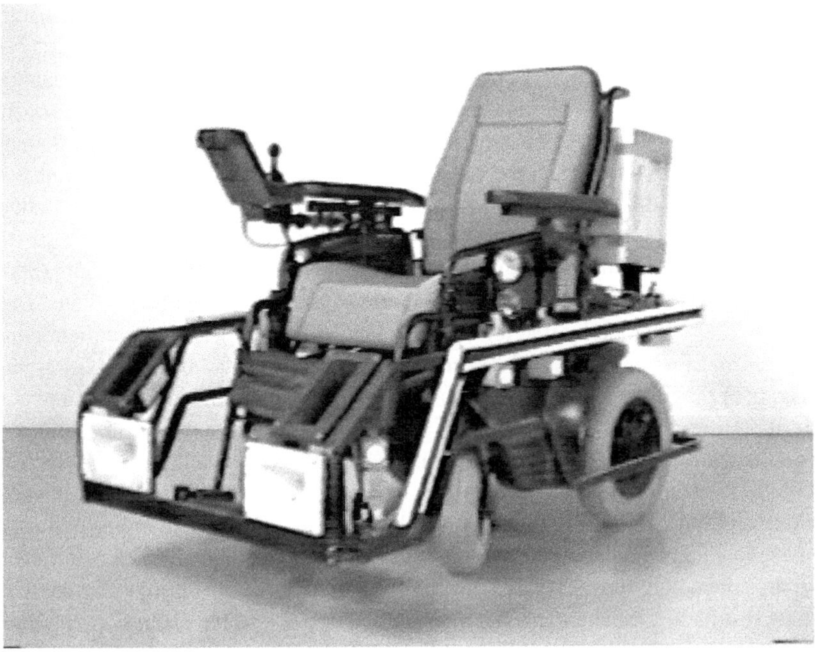

Figure 4.1 The SENARIO wheelchair platform. (Image adapted from Katevas et al., 1997 [177].)

of the powered wheelchair using the range data and a map of the room. The team treated the localization problem as an image processing problem, with each possible orientation and position of the robot in a room, represented by an *image* defined by the estimated lidar vector.

Thus, the localization process was reformulated as a parameter estimation problem, seeking to determine the vehicle model parameters (x, y position and orientation) that best fit the lidar range data obtained from the environment.[3] The complexity of the problem was compounded by transient signals generated by the movement of objects in lived home and hospital environments, resulting in significant transient noise in the lidar signal. To address this issue, the project employed stochastic diffusion search,[4] marking the first use of nature-inspired metaheuristic (NIM) with lidar in an AW environment perception system (see Section 7.3).[5]

4.3.1 SENARIO'S HYBRID ROUTE-PLANNING SOLUTION

The SENARIO AW used a unique hybrid navigation and path planner designed to operate in cluttered and dynamic environments such as hospitals and homes. Traditional navigation and path planning methods that relied on accurate metrical descriptions of the wheelchair environment were not suitable for this context.

Conversely, a purely reactive approach lacked the spatial modelling and reasoning capabilities required for goal-directed navigation.

To address this issue, the SENARIO hybrid navigation and global path-planning algorithms used *qualitative representations* constructed from the robot's sensor-aided interaction with the environment, along with a *topological map* of the domain, to structure route planning and the vehicle's movement. This approach integrated the live reactivity of direct perception through ultrasonic sensors with the centralized control offered by topological maps of the global environment.

Thus, instead of building a high-resolution global map of the environment, SENARIO's hybrid solution utilized qualitative representations directly constructed from the robot's sensor-aided interaction with the environment, effectively implementing a novel form of Rodney Brooks' *direct perception* reactive architecture. [48]

The path planner takes a line diagram of the robot environment as input and converts it into an enhanced grid, storing qualitative representations of variations in sensor behaviour between adjacent regions. An offline planner utilizes these representations to store motion commands at each grid cell, enabling the wheelchair to move in and out of rooms within a typical home environment. The online controller then accepts this enhanced grid along with the robot's starting and goal positions. It compares the actual sensor behaviour with the stored data in the grid, using the comparison results to estimate the robot's current position, retrieve planner instructions, and combine these instructions with risk-avoidance behaviours during navigation. [317]

Outside the specific constraints of AW operation in cluttered indoor institutional environments, more conventional localization and path-planning methods, which are better suited for real-time AV operation at high speeds, can be deployed. One such influential control architecture was Elann Gat's ATLANTIS (see Section 5.1). [112]

Notes

1. For a more detailed view of underlying technologies—besides sensors, also steer-by-wire, shift-by-wire, break-by-wire, V2X, 5G, HD maps—see, for example, McGrath 2019 or Chai 2021. [57, 237]
2. A video showing SENARIO undergoing user trials is linked here: https://tinyurl.com/6nj4dk3a
3. In environments that generate symmetric lidar data, this approach cannot *uniquely* resolve wheelchair position and orientation. Thus, results must be supplemented with approximate positional data obtained from other sensors.
4. First described by Bishop in 1989, *stochastic diffusion search* was the world's first *swarm intelligence* algorithm, [30] predating Dorigo's first publication on *ant colony optimization* by two years. [93]
5. Given a noisy lidar signal, the localization system had to determine the most likely position and vehicle orientation that could have generated the signal. Localizing within an environment comprising 50 rooms, each averaging $3m^2$, and considering the speed limitations of desktop (x286) PC computation at the time, a simple convolutional approach processing approximately 100K positions/sec would have taken over a hundred years to return just one position. In contrast, even on the relatively slow x286 PC hardware, stochastic diffusion search was able to return a ground truth location in seconds. This information was subsequently fused with higher-frequency odometer data to provide real-time position and orientation estimations as required.

5 The DARPA Grand Challenge for Autonomous Vehicles

In the pursuit of self-driving vehicles that could engage in battle or search and rescue missions, the US military research organization called the 'Defense Advanced Research Projects Agency' (DARPA) supported the development of next-generation AV techniques in the 2000s. DARPA aimed to address the challenge of developing ground vehicles, which, at the time, were less advanced than jet fighters or submarines that could already rely on computational support.

The development of AVs for ground transportation was a very complex task. In contrast to aircraft, which can fly above or around problematic air conditions, AVs require the ability to navigate unpredictable terrain and avoid obstacles such as trees, rocks, and buildings. Furthermore, they must identify *transient* obstacles, such as people and other vehicles, and consider how these might behave and how their movement could impact subsequent plans. [77]

As progress in AV development was plodding—measured against the US Congress's mandate that a third of military ground vehicles be unmanned by 2015—DARPA decided to conduct a challenge in 2004 and offer a reward of one million dollars to the team that could conclude the course first.[1] The challenge was the brainchild of its relatively new Director, *Anthony Tether*. This unconventional idea ultimately proved successful, eventually making Tether famous and laying the groundwork for a new AV ecosystem.

The approach of awarding prize money instead of employment contracts for research staff was innovative and had potential. If secret technologies, e.g. stealth aircraft or novel weapon systems, were involved, closed collaboration agreements like those of defence industry contractors would have been the right choice. However, the technologies typically deployed in AV design were not classified, which made open and public competition possible and defined a radically new way of innovation sourcing from universities, private companies, and individual contributors.

The DARPA Challenges began with competitor vehicles controlled by complex sets of manually configured rules. As the challenges became more complicated, such rule sets became increasingly difficult to define and maintain. By the time of the third Urban Challenge, control of the best vehicles had, in the main, shifted from rule-based to adaptive machine learning technologies, which proved a critical turning point in the development of autonomous vehicles.

DOI: 10.1201/9781003561460-7

5.1 DISCOVERING ATLANTIS

Extending the use of combined reactive and deliberative control, as explored in the SENARIO autonomous wheelchair, ATLANTIS was a three-layer AV control architecture proposed by Google engineer Erann Gat in 1998. [112] ATLANTIS addressed the limitations of previous approaches to AV control, such as *sense–plan–act* and Brooks' *subsumption architecture*, by integrating both reactive and deliberative elements within a new, three-layer architecture designed for dynamic, real-world scenarios, albeit at the cost of significantly increased complexity and computational demands. It is a heterogeneous, asynchronous system designed for controlling AVs based on a continuous action model and comprises of three main components, each with distinct roles.[2]

At its foundation is the reactive *controller* layer, which operates in real-time to handle immediate responses to environmental stimuli. The controller manages reactive processes, specifically primitive activities that do not require decision-making computations, and it focuses solely on the present moment, executing quick, reflexive actions. It directly connects to vehicle actuators, requiring high-speed operation to ensure swift reactions to changes in surroundings, such as obstacle avoidance.

Above the reactive layer is the *sequencer* layer, which acts as an intermediary between reactive and deliberative processes. This layer manages sequences of actions based on past experiences and current goals, coordinating actions over short periods to enable more complex behaviours. Doing so provides a balance between immediate reactions and more thoughtful, deliberative responses, enhancing the vehicle's adaptability.

The third module is the *deliberator* layer, which is responsible for long-term planning and decision-making. This layer utilizes a comprehensive *understanding* of the environment and vehicle goals to perform complex computations and determine optimal courses of action. It enables informed decisions aligned with overall objectives, such as finding efficient routes or planning manoeuvres to avoid potential hazards.

The ATLANTIS architecture allows AVs to adapt to dynamic environments. It significantly enhanced perception capabilities, enabling the distinction between stationary and moving objects, modelling object physics, including kinematic constraints, and tracking object movements over time. By integrating these layers, ATLANTIS enabled AVs to balance immediate reactions with longer-term planning, enhancing their efficiency and adaptability in complex, dynamic environments. To understand how such methods subsequently evolved, we will next review the rapid progress in AV technology that resulted from the first DARPA challenge.

5.2 THE FIRST DARPA (DESERT) GRAND CHALLENGE

The first DARPA Grand Challenge in 2004 marked the first time autonomous cars were tested in the real world, albeit under still highly restricted conditions in the Mojave Desert. Computers were twenty-five times faster than those available for Ernst Dickmanns' pioneering research almost two decades earlier. One hundred and six teams applied, twenty-five were down-selected, and fifteen made it to the starting line. The course was revealed to competitors only two hours before the race. All contestants received a CD with 2000 waypoints that could be located via onboard GPS equipment. [113]

The competition featured a variety of vehicles utilizing different technologies, including a military truck, a Humvee, and even a small motorcycle. While some cars were able to follow GPS waypoints accurately, they struggled with obstacle detection and collision avoidance, resulting in collisions during qualification rounds. Other vehicles were adept at sensing obstacles but struggled to follow waypoints or would falsely detect or *hallucinate* obstacles that were not present. None of the vehicles were able to complete the course, with all failures occurring between miles 5 and 8, which constituted only around 5% of the total course.

The DARPA report [365] outlines the specifics of these failures:

- *'**Vehicle 22 Red Team** (Carnegie Mellon; led by William Red Whittaker and Chris Urmson): At mile 7.4, the vehicle veered off course on switchbacks in a mountainous section, became lodged on a berm, and caught fire on the front wheels. The fire was quickly extinguished, and the vehicle was disabled by command.'*
- *'**Vehicle 21 SciAutonics II**: At mile 6.7, the vehicle became stuck in an embankment two-thirds of the way up Daggett Ridge. The vehicle was disabled by command.'*
- *'**Vehicle 9 The Golem Group**: At mile 5.2, the vehicle came to a stop on a steep hill, with the engine running and in gear but without enough throttle to climb the hill. After 50 minutes of trying, the vehicle was disabled by command.'*

In the competition, contestants predominantly relied on specialized hardware and engineering to overcome challenges. For example, speed adjustments were made by changing the voltage applied to the electric engine. To regulate engine speed under different load conditions, proportionalintegralderivative (PID) systems were deployed—feedback control technology with an engineering lineage that can be traced back to the 18th-century Watt governor (described on page 19).

In the DARPA competition, participants were required to navigate through challenging terrain by following a precise path. Although GPS coordinates were provided to all participants, this data was insufficient to plan a safe route across the terrain. Vehicles would need to integrate information from multiple sensors to build a coherent real-time representation of the local environment.

The Red Team—named after its frontman William Red Whittaker—aimed to complete the course by averaging 15 miles per hour. As their vehicle's sensors could not reliably look far enough ahead, the team leveraged their significant funding to gain an edge by creating a very detailed map of the competition course. The team believed that maps were fundamental to the effective operation of automated driving systems. Although manually annotating a map is costly, the approach looked to be effective, as terrain changes over time tend to be minor compared to the initial mapping and annotation efforts.

Sean Gerrish, an AI researcher at Google, describes how the Red Team created a highly detailed pre-race map, which they considered 'the best map of the world', to augment the one received on race day. [113] Leading up to the competition, they analyzed 54,000 square miles of the desert through satellite images to identify obstacles. In the two-hour pre-race window, fourteen workers manually annotated the route using several dozen computers. A computer continuously computed the best path, updating workers to prioritize areas of the map, and the final route was uploaded to their self-driving Humvee, dubbed *Sandstorm*, just before the race.

The Humvee determined its position on the map using a GPS sensor, although GPS data could have gaps and could not provide orientation/pose data. To address this, Sandstorm utilized the off-the-shelf, *Applanix POS-LV*, pose estimation system. This system combines inertial data with differential GPS signals via a Kalman filter, a groundbreaking method from 1960, developed by Rudolf Emil Kálmán, that integrates various measurements to estimate position over time. [171] Even during GPS outages, the Kalman filter could estimate position and velocity with only a few centimetres of error. This provided sub-metre accuracy even during extended GPS drop-outs. Additionally, it delivered precise angular data using carrier differencing from its pair of GPS antennas and inertial sensors.

To plan the most efficient route, the Red Team ingeniously combined pre-planned routes, calculated using their *best map of the world*, with real-time local environment data obtained from what Sean Gerrish vividly describes as a gigantic eye.

The *gigantic eye* was designed to scan the ground in the Humvee's path to detect obstacles not already mapped in the pre-planned route. The unit used lidar laser range-finding technology alongside two cameras, which were mounted on a gimbal for directional flexibility. While the cameras captured visual data, the lidar was busy creating a detailed point cloud representing the vehicle's local surroundings. This hybrid approach—*the best map of the world* combined with a *gigantic eye*—was designed to enable Sandstorm to navigate both known and unknown environments with a high degree of adaptability and safety.

The navigation process begins with a pre-planned path, which serves as a baseline for the vehicle's intended route. As Sandstorm moved, it continuously gathered data about its surroundings using its array of laser scanners. These sensors, including a long-range Riegl laser and short-range SICK scanners, provide a comprehensive view of the terrain ahead and any potential obstacles.

At the heart of Sandstorm's navigation capabilities are its *terrain evaluation* and *path adjustment* algorithms. For each laser scan, the system analyzes the safety of the terrain based on three key factors: the *steepness* of slopes, the *irregularity* of the surface, and the slope's *orientation* relative to the vehicle's direction of travel. This analysis results in a safety score for each point in the scan, creating a detailed map of traversable and potentially hazardous areas.

Once the terrain is evaluated, the system clusters the scan points into regions and identifies safe areas within the planned route corridor. The algorithm then calculates the centre points of these safe terrain clusters and measures their distance from the pre-planned path. Using sophisticated feedback control similar to that deployed in the vehicle's velocity control, the system smoothly adjusts the planned path to favour these safe areas while maintaining overall route integrity.

An essential feature of the path adjustment process is its forward-looking nature. Adjustments are only made further along the route than any previous modifications, ensuring a consistent and predictable navigation behaviour. This approach enabled Sandstorm to anticipate and react to terrain changes well in advance, resulting in smoother and safer navigation.

In addition to terrain evaluation, the system incorporates an obstacle detection mechanism using short-range laser scanners mounted on the vehicle's front bumper. This feature is handy in identifying immediate threats that the longer-range terrain evaluation might not capture. While primarily designed for use only in the qualification phase of the race, it added an extra layer of safety to Sandstorm's navigation capabilities.

The final step in the navigation process involves a path-tracking algorithm that follows the adjusted route while minimizing deviation. This algorithm translates the planned path into specific curvature and velocity commands, which are then executed by the vehicle's control systems.

As William Red Whittaker described in his race log, Sandstorm reached a maximum speed of 36 mph (57 km/h) and an average speed of 15 mph (24 km/h). Nonetheless, at a hairpin curve 7.4 miles (12 km) from the starting line in the Daggett Mountain area, it veered off the dirt road.

Whittaker assessed that this was likely due to several factors: a navigation algorithm error from sensor data, a poorly generated curve by the mapping path planner, a slow scan rate from one of the laser scanners, a GPS offset, or difficulty handling tight corridor constraints. As a result, the vehicle became high-centred on a berm—a raised grass strip/path beside the road—and could not move. Smoke from the car was subsequently identified as a rubber fire caused by the spinning wheels. This accident could have been avoided by using additional ultrasonic safety sensors.[3]

In retrospect, the Red Team's vehicle had all the elements of a modern AV system, albeit the rule-based vision system that analyzed local terrain proved to be its Achilles' heel, as it spectacularly failed to detect the nearby berm on which Sandstorm eventually became stuck.

As no vehicle came even close to finishing the race, the media spoke of a 'debacle in the desert'. [150] Hence, for Anthony Tether, the hoped-for breakthrough remained some time away. However, the starting point had now been set, and Tony was not about to lay down his arms just yet.

Let's recap: The underlying assumption of symbolic AI is that all (necessary) knowledge is explicit—or at least, to a significant extent, explicable in terms of symbols and symbol structures. Early researchers clearly recognised the necessity of a common body of knowledge—and referred to it as *common sense*—but they overlooked the fact that this core knowledge is mainly implicit or tacit.[4]

Engineers design structured rule sets to encode knowledge within well-defined domains. In essence, humans establish the framework of expertise while machines explore its specific applications. Algorithms operate as explicit, step-by-step guides for decision-making, introducing a symbolic layer where representations stand in for real-world entities. Beyond these tangible elements, symbols can be freely manipulated and processed, often through deductive reasoning and logical operations. This rule-based, handcrafted reasoning follows a linear, descriptive approach, producing declarative knowledge. Such a method aligns well with advanced cognitive tasks, such as mathematical analysis, which rely heavily on symbols and logic. While symbolic AI systems excel at reasoning and exhibit some perceptual abilities, they typically lack autonomous learning, making them less adept at abstracting knowledge from the world, and deal with uncertainty relatively poorly.[5]

The strength of the rule-based approach is that it excels in modelling hierarchy and sequential order, enabling logical inference. Its transparency allows for operation even on sparse data sets. Architecturally, it maintains a clear distinction between hardware and software, as well as between central processing and memory. However, its main drawback lies in its inability to effectively capture the complexity of real-world relationships, its lack of adaptability, and its inefficiency in leveraging computational power for learning and dynamic problem-solving.

Notes

1. In the subsequent narrative, the authors draw upon excellent descriptions of the DARPA Challenges by Sean Gerrish, Meredith Broussard, Alex Davies, and Lawrence D. Burns.

2. The ATLANTIS action model is based on operators that execute almost instantaneously. Rather than directly causing changes in the world, these operators initiate processes, referred to as activities, which then effect change. The operators responsible for starting and stopping these activities are called decisions. Decisions can initiate activities that, in turn, include computational processes capable of triggering additional activities. Provided there are no circular dependencies in this chain, activities can be organized into a hierarchy. At the top are high-level activities that involve computational processes, which initiate lower-level activities. At the base of this hierarchy are primitive activities, which are purely reactive sensorimotor processes without any decision-making computations.

3. The SENARIO robot wheelchair had dedicated hardware sensors in place to detect unexpected changes in height; for the SENARIO AW, these changes could potentially signal a stair. Hence, these sensors were able to generate emergency interrupts to the control system, halting the vehicle.

4. The operating principle of such cognitive processes was highlighted by Daniel Kahnemann and his distinction between *fast* or *abstract* (System 1) and *slow* (System 2) reasoning. System 1 reasoning happens mostly unknowingly, quickly, and often automatically. Obviously, the handcrafted rules of logical AI could only deal with Level–2 reasoning processes. This was the central limitation of the approach.

5. This is the summary of DARPA director John Launchbury; for AVs, he added: 'However, [Symbolic AI] does stumble when it comes to the natural world, and we saw this when DARPA was doing a lot of work in self-driving cars back in 2004 and 2005. We set a challenge to the community that was developing the technology for self-driving cars to say how can you put this technology together and drive 150 miles in the desert of California, Nevada. In 2004 none of the cars completed the course. They all failed; in fact, no car went further than eight miles down the course. And the problem was, they found it very difficult in the vision systems of the car to be able to distinguish a black shape in the distance whether it was a shadow or a rock, should I avoid it? Should I barrel through it? And they kept making the wrong decisions. A year later, in 2005, we had very many successful cars. In fact, five cars completed the whole course. A big change was that they started to use machine learning. They started to use techniques that were probabilistic in the way that they handled information.' [196]

Section III

Data and Learning

If we identify the *first wave of vehicular autonomy* as one driven by the intelligent application of symbols and rules, the *second wave of vehicular autonomy* was to be characterized by the application of statistics and mathematical machine learning.[1]

For connectionists, many tasks involving intelligence, such as perceiving, listening, and reading, are performed subconsciously in a subsymbolic form. Unlike symbolists, who explicitly define decision-making processes, connectionists automate these tasks using [statistical] models trained on big data, replacing symbols with distributed mathematical operations.

Their typical application centres on pattern identification, learning from input and output data without the use of predefined rules. These models are adaptive and automate non-formalizable tasks, but are inherently opaque (black box).

Machine learning (ML) was a term introduced in 1959 by Arthur Samuel during his pioneering work on creating computer programs that could improve their performance through experience. [296] Samuel defined ML as 'the ability of computers to learn from experience and improve automatically', laying the foundation for what would become a key branch of artificial intelligence. Rather than relying on explicit programming to perform tasks, ML focuses on developing statistical algorithms capable of learning from data. Once trained, these programs aim to generalize to new, unseen data, enabling them to handle novel tasks without explicit instruction.

Samuel's work specifically involved programming a computer to play the game of draughts (checkers) at an increasingly skilled level. The program was designed to learn from its gameplay through techniques such as self-play and the statistical evaluation of board positions. This groundbreaking approach, which in principle also forms the basis of much more powerful modern systems such as AlphaZero, [312] demonstrated that computers could perform tasks without being explicitly programmed for every possible scenario. Instead, the system autonomously improved by analysing past outcomes and refining its strategies.

While machine learning is today most often associated with the connectionist school of AI, it is essential to recognize that it can also be effectively applied in symbolic AI (GOFAI) systems, such as Samuel's checkers system and neuro-symbolic systems.

At the time of writing, companies such as Google, Amazon, and Facebook, which have access to vast amounts of data, are playing a dominant role in the development of machine learning technology. They have made significant breakthroughs in ML for vision by creating large labelled datasets, such as ImageNet (which contains 14 million pictures of 22,000 classes). These datasets have been crucial for advancing the field, enabling computers to make accurate predictions based on subtle patterns in data.

6 Forms of Machine Learning

There are three fundamental forms of machine learning contingent on the way they learn from data: (i) supervised learning, (ii) unsupervised learning, and (iii) reinforcement learning. The key difference between them lies in the way they process human input and feedback. To understand how ML systems are deployed in practice, it is first necessary to sketch these approaches.

6.1 SUPERVISED LEARNING

Supervised learning is one of the most popular types of ML used today. It is a form of machine learning in which a *model* is trained using a *training set* of data that contains both *known inputs* and *corresponding known outputs (labels)*. The objective of ML is for the model to learn to encapsulate the relationship between the inputs and outputs, enabling it to correctly *predict* output labels for new, unseen input data.

For example, suppose a supervised ML system is being trained to recognize dogs in pictures. In that case, a large number of carefully labelled images of dogs need to be provided as training set data, and the ML system's output will indicate whether a given input contains a dog or not.

In use, the ML model is iteratively updated (*trained*) by minimizing the difference between its predictions and the actual labels, typically using mathematical techniques such as gradient descent to adjust its parameters and reduce its prediction error.[2] The process of presenting to the system all the input-output pairs defined in the training set is defined as one *epoch* of learning.

Thus, the typical supervised learning process involves (i) collecting labelled data, (ii) training the model on this data, and at regular intervals, (iii) evaluating the model to see how it performs on a *validation* data-set.[3] This results in the calculation of the current system *prediction error*, which is subsequently used to tune the model's hyperparameters and prevent overfitting to the data.[4] When the model's prediction error starts to rise, training needs to stop as the model is losing the ability to *generalize*. Finally, (iv) the model's performance is evaluated on a *test set* of *unseen* input data.

Supervised learning is usually deployed as *predictive models* to address two main types of problems: *classification*, where the goal is to *predict* discrete labels, e.g. classifying images as containing dogs or not; classifying emails as spam or not, etc., and *regression*, which involves *predicting* continuous values, e.g. estimating house prices given a set of house features: location, number of rooms, etc. Standard algorithms used in supervised learning include linear regression, decision trees, support vector machines (SVMs), neural networks, etc., each of which we will subsequently examine in detail to gain a comprehensive understanding of the various approaches and the diversity of strategies.

DOI: 10.1201/9781003561460-9

Supervised learning is widely used in applications such as email filtering, image and speech recognition, natural language processing, predictive maintenance, and predictive analytics.

6.2 UNSUPERVISED LEARNING

In contrast to other forms of machine learning, in its conception, unsupervised learning is more akin to how humans typically learn. Rather than following predetermined instructions, humans start learning by simply observing and experiencing the world around them. This process is analogous to unsupervised learning, where the goal is to interpret data without using explicit labels; it is akin to looking for the proverbial *needle in a haystack* without having any predefined notion of what a needle looks like. The objective is to create a mapping to an output, or output space, without explicit guidance on what that output should be.

Since there are no labels to guide the learning process, the model tries to group, cluster, or reduce the dimensionality of the data based on its inherent similarities or differences. For instance, in *clustering*, the algorithm divides the data into groups (clusters) based on similarity. In contrast, in *dimensionality reduction*, the model simplifies the data by finding important features or compressing it into a lower-dimensional space.

The key distinction from supervised learning is that unsupervised learning works with unlabelled data, meaning the model must rely solely on the data's internal structure rather than being explicitly trained on known input-output pairs.

NB. In unsupervised learning, the use of a validation set to mitigate, say, overfitting is uncommon. This is because unsupervised learning does not involve labelled data, so there's no explicit ground truth data against which to validate the model predictions.

Historically, unsupervised learning was typically used for data pre-processing or transformation, with clustering—grouping or sorting data using a defined similarity metric—being one of the most common applications. Others include data compression (to reduce the *dimensionality* of data) and anomaly detection, where the system seeks to identify unusual patterns in the data, say, corresponding to suspicious user behaviour.

More recently, Masking provides a ground truth, SSL in my view is very different in this respect from unsupervised learning of unsupervised learning have proven very effective in training large language models, such as ChatGPT, on vast unlabelled text corpora scraped from the internet. Here, for example, a sequence of words is given as input and unsupervised learning models are used to predict the most likely next word in the sequence.

6.3 REINFORCEMENT LEARNING

The process of a player improving their performance in a computer game is a classic example of reinforcement learning (RL), a type of ML used to train machines via a form of *trial and error*. Unlike supervised learning, where the correct response to a known input is provided, reinforcement learning requires an *agent* to *interact* with the *environment* and receive feedback in the form of *rewards* or *penalties* to achieve a specific *goal*. To learn how to succeed, the agent must try different strategies and adjust its actions based on the outcomes of each approach.

Thus, the agent at the core of reinforcement learning systems is actively influencing the environment through its *interaction* rather than merely observing it. This method of reinforcement has roots in the work of B.F. Skinner's *operant conditioning* and the behaviourist school of psychology.[5][326] The agent seeks to find the right balance between *exploration* of the total search space of solutions and *exploitation* of potential solutions that current evidence suggests might be good. In this way, the agent learns from both positive and negative feedback to improve its performance over time.

Reinforcement learning is best suited for well-defined problems with finite contexts, such as computer games, where the rules provide constraints for the agent to operate within. For example, DeepMind used reinforcement learning as a core component of AlphaGo, an AI system that defeated the world champion of the board game Go (see page 25). [322]

Conversely, in many *real-world* tasks, where the environment is open and infinite, reinforcement learning can become data-hungry and require countless trials for successful learning. Indeed, as Yann LeCun wryly observed, training a self-driving car via reinforcement alone would require driving millions of hours, harming thousands of pedestrians and driving off a cliff multiple times before it eventually figures out that this is not such a great idea ...

In contrast, recent experiments combining reinforcement learning with imitation learning (IL) (a general class of algorithms that enable learning directly from demonstrations) have begun to show promise in some complex autonomous driving scenarios. [41]

Combined imitation and reinforcement learning (CIMRL) seeks to integrate the strengths of imitation learning and reinforcement learning to overcome their respective limitations. While IL is prone to compounding errors in unfamiliar scenarios, RL often struggles to define human-like driving behaviours through reward functions alone.

CIMRL addresses these challenges by employing multiple trajectory generators, including IL-based models, to propose potential driving plans. An RL-based policy then selects the safest and most effective course of action by evaluating these plans in a closed-loop manner. This evaluation considers both immediate and long-term outcomes, ultimately choosing the plan with the highest score while avoiding potential risks. The effectiveness of CIMRL has already been demonstrated in simulations, such as the Waymax simulator, as well as in some real-world testing. [265]

Notes

1. For the surprisingly simple maths behind all deep learning techniques, see Ananthaswamy (2024). [9]
2. Exceptions to iterative, gradient descent learning include one-shot paradigms, such as Igor Aleksander's *weightless networks*; however, these are now not often deployed.

3. In machine learning, the validation data-set is a subset of the total available training data, used to tune the model's hyperparameters and evaluate its performance *during* training.

4. Model overfitting occurs when a machine learning model learns not just the underlying patterns in the training data but also the noise or random fluctuations that are irrelevant to the task. As a result, the model performs exceptionally well on the training data but fails to generalize to new, unseen data, leading to poor performance in real-world applications or on a validation/test set.

5. Reinforcement Learning and Operant Conditioning share many fundamental similarities, as both involve learning from the consequences of actions and adapting behaviour based on rewards or punishments: operant conditioning, a psychological theory developed by B.F. Skinner is one of the most direct inspirations for the development of reinforcement learning. However, while they are conceptually aligned, there are some key differences in their application and complexity.

7 Types of Machine-Learning Models

Various types of machine-learning (ML) models have been proposed that are contingent on the target application. Some models, such as artificial neural networks and deep learning, have found widespread application in autonomous vehicles; others, such as linear regression and evolutionary computing, are less widely used. Nevertheless, we believe it is essential to understand the historical developments in a broader context to comprehend the progress of the conceptual evolution that underlies autonomous driving technology.

7.1 REGRESSION

Regression methods in ML have their roots in statistics and were developed long before more modern forms of machine learning emerged. Linear regression describes a simple type of machine learning used to predict continuous outcomes. It models the relationship between a dependent variable (the target) and one or more independent variables (the features) by fitting a *linear* equation to the data. In its simplest form, this equation is $y = mx + b$, where y represents the predicted outcome, x is the feature, m is the slope (indicating the strength of the relationship), and b is the intercept. When multiple features are involved, it becomes a multivariate linear regression.

Linear regression is primarily used for regression tasks where the objective is to predict a numerical value. Its strengths lie in its simplicity and interpretability, as the coefficients in the linear equation directly indicate how each feature influences the prediction. Additionally, it is computationally efficient, making it suitable for problems with straightforward linear relationships.

To see how linear regression works, imagine you own a shoe store and wish to predict a person's shoe size based on their height. After collecting data, you observe that taller people tend to have larger feet. To make recommendations easier, you decide to build a simple linear model to estimate shoe size from height.

To achieve this, you plot the data and draw a straight line through the points, representing the relationship between height (input) and shoe size (output). The aim is to find the line that best fits the data. The best model of the data will minimize the difference between the predicted and actual shoe size. This is the essence of linear regression: It finds the line that predicts the outcome (shoe size) from the input (height) by minimizing prediction errors.

In this analogy, height is the input feature, shoe size is the target to predict, and the line represents the model that captures the relationship between the two. The process of adjusting the line to fit the data mirrors how linear regression minimizes errors to find the optimal prediction model.

DOI: 10.1201/9781003561460-10

Although linear regression is used to model a linear relationship by default, non-linear extensions have been developed to allow it to capture more complex patterns in the data. Two widely used techniques to extend linear regression to handle non-linear equations are:

Basis functions: In this approach, the original input features are transformed into new, non-linear features using *basis functions*. The basis function maps the input data into a higher-dimensional space where a linear model can better fit the non-linear patterns in the original data. Some standard basis functions include *polynomial basis*—where, although the model remains linear in the parameters, i.e. the coefficients, the transformation of input features into polynomial terms enables it to model non-linear patterns—and radial basis functions (RBFs), which will be discussed later.

Generalized linear models (GLMs): Here, the linear regression model is extended by using a link function that transforms its output into a non-linear form. For example, in logistic regression (a type of GLM), the output is transformed using the logistic function to model probabilities.

7.2 DECISION TREES

Decision Trees are a popular *symbolic* ML paradigm used for both classification and regression tasks. They work by creating a tree-like model of decisions based on features in the data, essentially breaking down a data set into smaller subsets while incrementally developing an associated Decision Tree.

The tree structure consists of a root node, internal nodes, and leaf nodes. Each internal *node* represents a *test* on an attribute, e.g. whether a coin flip comes up heads or tails, each *branch* represents the *outcome* of the test. Each *leaf* node represents a *class* label or a numerical value (in the case of regression). The path from root to leaf represents classification rules.

During training, the tree is constructed top-down, and the best attribute is chosen to split the data at each step. This choice is typically made using metrics such as *Gini impurity*,[1] or *information gain*.[2] [45] The process continues recursively on each subset until a stopping criterion is met, e.g. all samples in a node belong to the same class or the tree reaches a maximum depth. To make predictions, new data is fed through the tree, following the appropriate branches based on its features, until it reaches a leaf node, which provides the prediction.

To understand how Decision Trees are applied in ML, imagine you are house-hunting with a list of criteria, such as price, number of bedrooms, whether it has a garden and its distance to work. You begin by evaluating the most crucial factor, such as price. If the house is affordable, you then consider the next factor, like the number of bedrooms, and continue until you either decide to buy the house or reject it.

This mirrors how a Decision Tree works in machine learning. The tree begins by making decisions based on the most informative feature, e.g. price, and continues to branch based on additional features, splitting the data at each step until a prediction is made. In this analogy, the features represent the criteria, the decisions correspond to splits in the tree, and the final decision to buy or not is the model's prediction. Like in house hunting, the Decision Tree evaluates features step by step to reach a conclusion.

7.3 NATURE-INSPIRED METAHEURISTICS

Nature-inspired metaheuristics (NiMs) are a class of ML models inspired by natural *phenomena*, typically used to solve complex optimization and learning problems. These approaches mimic *processes* observed in nature to search for optimal solutions in large, often multidimensional spaces.

NiMs are population-based, a form of *swarm intelligence*, wherein knowledge is encoded and learnt over time by adjustments to the parameters that define each agent in the population. Among these methods, several stand out for their effectiveness and widespread use. Examples include evolutionary strategies, particle swarm optimization, ant colony optimization and stochastic diffusion search.

Evolutionary strategies (ES) simulate natural selection by starting with a population of candidate solutions, where each individual's genes represent a potential solution to the problem. Over generations, better solutions emerge as the most successful individuals reproduce and pass on their traits. This process, driven by mutation, recombination, and selection, modifies the genotypes of the population, progressively increasing its overall fitness over generations.

Particle swarm optimization (PSO), inspired by the social behaviour of bird flocking or fish schooling, uses a population of particles that move through the solution space of the problem, keeping a certain distance from their neighbours and adjusting their trajectories based on their own best-known position and the best-known position of the entire swarm.[3]

Ant colony optimization (ACO) is a method inspired by how real ants find the shortest path to food. In nature, ants leave behind a chemical trail, known as pheromones, as they move. Other ants follow these trails, and the more that ants use a path, the stronger the pheromone trail becomes, making it more attractive to others. Over time, the ants converge on the shortest route to the food because that path accumulates the most pheromone.

Stochastic diffusion search (SDS), inspired by the way social insects collaborate to find resources in nature, employs a population of agents that interact and exchange information to locate optimal solutions in the search space. [30] In SDS, agents perform cheap, *partial evaluations* of a hypothesis (a candidate solution to the problem). Agents then share information about hypotheses (diffusion of information) through direct one-to-one communication. As a result of this diffusion mechanism, high-quality solutions can be identified from clusters of agents maintaining the same hypothesis.

NIMs have proven particularly useful for problems where traditional gradient-based optimization techniques may struggle, such as those with many local optima or discontinuous cost functions. For example, in the context of our exploration of autonomous vehicles, we have already observed how SDS has been successfully used in autonomous wheelchair localization (see page 59).

7.4 ARTIFICIAL NEURAL NETWORKS

Artificial neural networks (ANNs) are a class of ML models inspired by a specific class of natural *entities*—the structure and function of the human brain. They consist of layers of interconnected nodes—*artificial neurons*—that process data and learn patterns through training on large data sets. The information they process is not represented by **symbols** but by **tensors,**[4] patterns of activation, in which inputs and outputs to the network are typically represented as a *vector* of *real-valued* numbers.

Today, it is common to refer to large artificial neural networks by the moniker *deep learning*, a term which was introduced into the field of machine learning by Rina Dechter in 1986 [80] and explicitly applied to ANNs by Aizenberg et al. in 2000. [2]

In its broadest sense, neural computing is a form of ML, a mode of problem-solving based on learning from experience as opposed to classical computational, syntactically specified, *algorithmic* methods. At its core is the study of networks of adaptable nodes which, '*through a process of learning from task examples, store experiential knowledge and make it available for use.*' [6] In this sense, an ANN is constructed by appropriately connecting a set of adaptable nodes, and so construed, also encompasses *kernel nodes*, as deployed in *radial basis function networks*. [49]

- A *single-layer neural network* only has one layer of adaptable nodes between the input vector, X and the output vector O, such that the output of each of the adaptable nodes defines one element of the network output vector O.
- A *multi-layer neural network* has one or more hidden layers of adaptable nodes between the input vector and the network output. In each of the network *hidden layers*, the outputs of the adaptable nodes connect to one or more inputs of the nodes in subsequent layers and in the network *output layer*, the output of each of the adaptable nodes defines one element of the network output vector O.[5]
- A *recurrent neural network* is a network where the output of one or more nodes is fed back to the input of other nodes in the architecture, such that the connections between nodes form a directed graph along a temporal sequence, so enabling a recurrent network to exhibit temporal dynamics; enabling a recurrent network to be sensitive to particular *sequences* of input vectors.

Since 1943, various frameworks for adaptable nodes have been proposed.[6] However, the most common—as currently deployed in deep neural networks (DNNs), generative AI and large language models—remain grounded on the generalized McCulloch–Pitts model (see Figure 2.1). [236]

A simple analogy illustrates the core ideas behind neural computing. Imagine a factory where raw materials are processed by a series of workers, each performing a specific task and passing the product to the next worker. If the final product is defective, feedback is sent back through the line, allowing each worker to adjust and improve their actions. Over time, the workers learn to process the materials more effectively, producing high-quality products.

This mirrors how artificial neural networks operate. The raw materials represent the input data, and the workers are the neurons, each performing a small computation. The final product is the network's prediction. When the prediction is wrong, presenting feedback to the neurons helps them adjust their settings (weights). Over time, the network improves its predictions by learning from this feedback process. In this analogy, the network, like the factory, learns to process input data more efficiently through iterative feedback, gradually improving its performance.

7.4.1 THE SINGLE-LAYER PERCEPTRON

An influential figure in the early development of neural computing was Marvin Minsky, who published his doctoral dissertation on neural networks in 1954. [243] Four years later, Frank Rosenblatt, a former classmate of Minsky, introduced his influential neuron-like computational model, *the perceptron*. [285] Rosenblatt's model incorporated both the fixed-weight and learning versions of McCulloch–Pitts neurons, with the adaptive synapses represented as variable multipliers of input signals, systematically adjusted to minimize error. Rosenblatt famously claimed that [in the Class C perceptron] '*for the first time we have a machine capable of having original ideas*'.

A few years later, Rosenblatt published his *learning rule* for perceptrons the *perceptron convergence procedure (PCP)* in his 1962 monograph, *The Principles of Neurodynamics*, where, by learning weights automatically, the PCP was regarded as a significant advancement in neural networks. [286] Indeed, at the time, it was widely claimed that the perceptron was beginning to reveal the very rules that drive brain mechanisms' and attracting considerable media attention for its brain-like' qualities.

The single-layer perceptron (SLP) model can be divided into three layers (see Figure 7.1). The first layer is the retina of the perceptron, consisting of a regular array of *Sensory Units* (S-Units). The second layer, made up of *Association Units* (A-Units), receives input as the weighted sum of outputs from selected S-Units. These weights are fixed, meaning A-Units act as feature detectors, responding selectively to patterns on the retina. The third layer, comprising *Response Units* (R-Units), has variable-weighted connections to specific A-Units. An R-Unit outputs $[+1]$ if its weighted input sum exceeds a threshold T, and $[-1]$ otherwise. In some models, an active R-Unit inhibits all non-connected A-Units.

7.4.1.1 The perceptron convergence procedure

In the perceptron convergence procedure, Rosenblatt proposed that if the perceptron's response to an input vector is correct, no change is made to the R-Unit weights. If the response is incorrect, the procedure either decrements all active weights and increases the threshold if the R-Unit fires when it should not, or increments active weights and reduces the threshold if the R-Unit fails to fire when it should.[7]

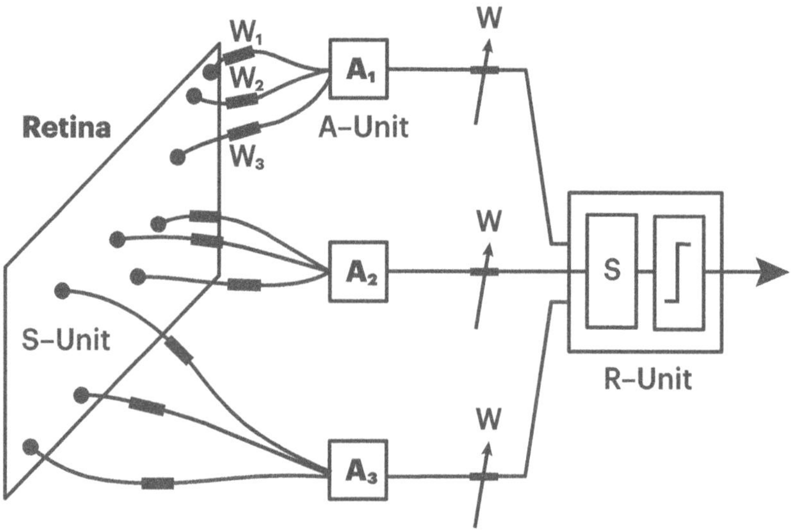

Figure 7.1 The structure of Rosenblatt's perceptron. (Image adapted from Bishop, 1989 [32].)

7.4.1.2 Perceptron convergence theorem

Rosenblatt's perceptron convergence theorem (PCT), based on a fixed increment approach, guarantees that the above procedure will find a set of weights that performs a specific mapping in a single-layer perceptron, *provided such a set exists*—i.e. if the problem is linearly separable, by which we mean that a line, or in general a *hyperplane*, can be constructed to partition the classes correctly.

7.4.2 THE SIMPLE DELTA (OR WIDROW-HOFF) LEARNING RULE

In 1960, Bernard Widrow and Marcian Hoff of Stanford introduced an adaptive learning model called ADALINE (ADAptive LInear NEuron), which they subsequently refined as MADALINE (Multiple ADAptive LINear Elements). [377]

ADALINE deployed a linear version of the McCulloch–Pitts neuron during the *learning* process but used a *Heaviside* (threshold) activation function to produce a binary output for *classification*. Thus, during *training*, the output from the linear unit is not passed through the Heaviside function; instead, given an input vector and a weight matrix, it calculates its output as the weighted sum of the input and compares this with the target output.

The ADALINE weight update process follows the simple delta rule (SDR), which adjusts weights based on the *error*—the difference between the neuron's *target value* and its *actual output*—with the adjustment to each weight proportional to the input multiplied by the error and the *learning rate*.

Since the learning rate controls the size of the update, it is a critical *hyperparameter* that affects the efficiency and accuracy of the learning process. However, for actual classification, a threshold function is applied to the weighted sum to produce a binary output, typically +1 or -1, depending on whether the net input is greater than or equal to a threshold or less than it.

ADALINE was initially developed for binary pattern classification. Its successor, MADALINE, was famously used as an adaptive filter to eliminate echo on phone lines and represents one of the earliest applications of an artificial neural network to a real-world engineering problem.

7.4.3 THE FIRST AI WINTER

While symbolic AI typically required handcrafted heuristic search algorithms, neural networks introduced algorithms that could iteratively adjust their parameters (effectively configuring themselves) by a process akin to human learning. Initially, both research in schools progressed and developed in parallel. Infamously, this methodological pluralism came to an abrupt stop in 1969 with the publication of Marvin Minsky and Seymour Papert's iconoclastic text 'Perceptrons: An Introduction to Computational Geometry'—a book which mathematically exposed fundamental limitations on the type of tasks that can be realized using single-layer neural networks. [244] For example, Minsky and Papert showed that perceptrons are unable to calculate such mathematical functions as *parity* or the topological function of *connectedness*.[8]

Henceforth, research into neural networks rapidly fell out of favour, and symbolic AI, instantiated using Physical Symbol Systems, emerged triumphant as the dominant paradigm for AI research.

However, that triumph did not last long, and the victory proved pyrrhic when, in 1973, Sir James Lighthill published his influential critique of artificial intelligence, foregrounding problems of combinatorial explosion. [211] Highlighting, for example, that if the *PSS solution generator* forms B branches, and D is the depth of search, then we have a search tree requiring, in the limit, B^D tests.

By way of illustration, the game of chess involves very broad trees, which rapidly describe enormously large search spaces, albeit human chess players, in using their intelligence and knowledge of the game, are highly selective in what branches to consider and thus evaluate a tiny fraction of this space.

Published in 1973, Lighthill's report was commissioned by the British Science Research Council (BSRC) as an evaluation of academic research in the field of artificial intelligence. The report gave an extremely pessimistic prognosis for many core aspects of research in this field, concluding that: '*In no part of the field have the discoveries made so far produced the major impact that was then promised.*'

Henceforth, although the BSRC continued to support AI research related to automation and computer simulation of neurophysiological and psychological processes, it was highly critical of basic research in foundational areas such as robotics and language processing.

Subsequently, the report formed the basis for 'the decision by the British government to end support for AI research in most British universities', and to a lesser extent, in the USA. [288] This lack of interest in AI—both symbolic and connectionist—persisted for a decade and is generally considered to mark the first so-called '*AI winter*'.

During the period 1969–1986, following the publication of Marvin Minsky and Seymour Papert's *perceptrons*, only a relatively small number of scientists continued their research into artificial neural networks (ANNs). These researchers formed the core of a small, international group that upheld the principles of connectionism for nearly two decades, during which the symbolic, rule-based school of AI held hegemony. These researchers included:

- Igor Aleksander, who, from the 1960s onwards, pioneered and developed the field of weightless/nTuple neural networks. [5]
- Stephen Grossberg, who developed adaptive resonance theory (ART) networks. [123]
- John Hopfield, who in 1982 published the mathematical foundations of Hopfield networks.[9] [151]
- Teuvo Kohonen, who in 1982 developed the *self-organising feature map* which, in the guise of the *phonetic typewriter*, found successful application in the nascent field of speech-to-text systems. [186]

7.4.4 THE MULTI-LAYER PERCEPTRON

In 1986, Hinton, working with Rumelhart, Williams, and the PDP Research Group at MIT,[10] announced the application of the backpropagation (BP) algorithm, also known as the generalized delta rule (GDR), to train multi-layer perceptrons (MLPs).[11][287] This development overcame many of the serious limitations of single-layer perceptrons identified in Minsky and Papert's influential 1969 critique. [244]

The development of the GDR was a seminal moment in neural network research, leading to a new renaissance in the field of ANNs in the 1980s. It defined how to propagate an error term backwards through the network—from the output units to the inputs—defining concomitant weight changes to the network structure such that the overall prediction error of the network is minimized.

Learning in a multi-layer perceptron network can be viewed as the process of curve fitting that occurs as the MLP adapts its structure to model the training data. If there are K (weights + biases) in the MLP network, then, over time, the GDR traces out a path in $(K+1)$ dimensional weight–error space via the application of the chain rule of calculus to propagate the error gradient backwards through the network. This allows for weight adjustments at each layer, effectively minimising prediction error across the entire network. If the learning process is effective, the network will converge to a local minimum in this weight–error space by the end of the learning process.

The subsequent explosive resurgence of interest in connectionism generated by backpropagation and MLPs took the AI and cognitive science community by surprise. For example, a 1987 conference on PDP methods held at the University of Oxford by the *Experimental Psychology Society* attracted an audience of 700 (including one of the authors of this work), more than double the anticipated maximum attendance. Whilst in San Diego, the first *IEEE International Conference on Neural Networks* attracted almost 2,000 participants. As Dreyfus and Dreyfus later remarked, it was as though '*Neural network modelling may be getting a deserved chance to fail, as did the symbolic approach.*' [95]

7.4.4.1 The application of multi-layer perceptrons

The early applications of both SLPs and MLPs were built on one of two basic functional architectures:

Regression architecture: A regression architecture is used for function approximation. An example of complex function approximation might be: given the weather and season, what is the likely demand for electricity in the UK?
In 1989, George Cybenko proved *the universal approximation theorem for multilayer perceptrons*. [74] Cybenko demonstrated that a feedforward neural network with a single hidden layer, using a sigmoid activation function, can approximate any continuous function on a closed and bounded interval, given sufficient neurons in the hidden layer. Kurt Hornik subsequently extended this work, showing that MLPs with various activation functions, not just sigmoidal ones, also possess this universal approximation property. [152]

Discriminative architecture: A discriminative architecture—*or discriminative classifier without a model*—can be used to *discriminate* (classify) the value of the target variable Y, given an observation x. An example of a discriminative task could be, 'Does a given input image contain a tank?'
Such experiments in machine classification were actually carried out and reported as long ago as 1964 in a paper by Kanal and Randall, 'Recognition system design based on statistical analysis'. [172] As Kanal subsequently describes in his 1991 reminiscences:

> 'In 1962–63 at Philco-Ford, seeking a systematic approach to designing layered classification nets, we decided to use a hierarchy of threshold logic units with a first layer of *feature logics* which were threshold logic units on overlapping receptive fields of the image,[12] feeding two additional levels of weighted threshold logic decision units. The weights in each level of the hierarchy were estimated using statistical methods rather than iterative training procedures.' [173]

Furthermore, in their 1964 paper, Kanal and Randall report outstanding results: '*The experimental performance of the statistical classification procedure exceeded all expectations.*' [172] Nonetheless, over the years, their work led to the emergence of a widespread *urban myth* about machine learning:

> 'The story goes that researchers trained the MLP on a data set of images, some containing tanks and others without. Initially, the model seemed to perform well, correctly identifying tanks in images. However, it was later discovered that the network had not learned to recognize tanks at all. Instead, it had learned to differentiate between photos based on the background conditions. The tank images were taken on cloudy days, while the non-tank photos were taken on sunny days. Thus, the network had learned to associate cloudiness with tanks rather than recognising the tanks themselves.'

In his 1991 reminiscences, Kanal emphasizes the success of the 1964 experiments, effectively refuting the urban legend, at least as it pertains to his work [173]. However, the myth—despite existing in several variants, some of which appear plausible—highlights a significant issue in machine learning: the need to understand data set biases and ensure that neural networks are learning the intended features. This challenge remains pivotal in contemporary AI and machine learning development, and the anecdote serves as a cautionary tale.

7.4.4.2 The drift away from multi-layer perceptrons

During the 1990s, however, after a decade in the sun, research in ML began to shift from multilayer perceptrons (MLPs) to support vector machines (SVMs) and Bayesian methods. At the time, training MLPs using backpropagation was slow and prone to issues such as local minima and overfitting, particularly on small or noisy data sets. Furthermore, MLPs lacked the theoretical guarantees that other methods provided, making them more challenging to interpret and trust. Additionally, regularization techniques were not as advanced, which further contributed to their poor generalization.

In contrast, training algorithms for SVMs and many Bayesian methods were faster and more stable than those for MLPs. Given the limited computational power and data available in the 1990s, these methods were much more suitable for practical applications.

7.4.5 THE SUPPORT VECTOR MACHINE

The support vector machine (SVM), introduced by Cortes and Vapnik in 1995, [71] building on Vapnik's earlier foundations, [355] addresses a number of the limitations of MLPs. They are predominantly used for classification tasks, where they attempt to find the optimal hyperplane that separates different classes of data points with the *maximum margin*, leading to better generalization, especially in high-dimensional spaces. However, SVMs can also be applied to regression problems through a variant known as support vector regression (SVR). [96]

In its simplest form, an SVM works for binary classification by plotting data points in a high-dimensional space and identifying the optimal hyperplane to separate the two classes. The *best* hyperplane is the one that maximizes the margin, which is the distance between the hyperplane and the nearest data points from each class. These nearest points are called *support vectors*, giving the algorithm its name.

For data that isn't linearly separable, SVMs use a technique called the *kernel trick* to transform the input space into a higher-dimensional feature space where a separating hyperplane can be found without significantly increasing computational complexity. Put simply, imagine a set of points on a flat surface that can't be divided by a straight line. The kernel trick is akin to mapping these points into a 3D space, where you can now separate them with a flat plane. However, instead of actually moving the points (which would be computationally expensive), the kernel trick provides a way to calculate how the points would behave in this higher dimension without relocation.

This process is somewhat akin to solving a puzzle by looking at its reflection in a curved mirror—the reflection reveals a solution that wasn't obvious when looking at the puzzle directly. The kernel function acts like this curved mirror, unveiling patterns in the data that weren't visible in the original space and allowing the SVM to find separations that weren't possible before.

The use of convex optimization in SVMs ensures that the global minimum is found, avoiding the convergence problems faced by MLPs. SVMs also have a strong theoretical foundation, with concepts such as the Vapnik-Chervonenkis dimension offering insights into model performance. [355] Furthermore, the kernel trick allowed SVMs to handle both linear and non-linear problems effectively, making them a versatile ML solution for a wide range of applications.

7.4.6 BAYESIAN METHODS

Bayesian methods offer another alternative to multi-layer perceptrons. Their probabilistic framework offers a sophisticated yet practical approach to handling uncertainty and making predictions. Named after the 18th-century statistician Thomas Bayes, these iconoclastic techniques revolve around updating beliefs as new information becomes available and represent a sharp opposition to so-called Fisherian statistics—also known as frequentism—which still leads to fierce controversy among statisticians today.[13]

At its core, Bayesian learning starts with an initial set of assumptions or educated guesses about a problem. As new data is acquired, these initial beliefs are systematically adjusted, much like how we refine our opinions after learning new facts. The strength of Bayesian methods lies in their ability to combine prior knowledge with fresh data in a mathematically rigorous way. This approach yields predictions that are not just single values, but probability distributions, providing insight into both potential outcomes and the confidence in those predictions. This nuanced view of uncertainty is particularly valuable for decision-making in complex, real-world scenarios.

For example, imagine you are a detective trying to solve a mystery, and each piece of evidence you gather—like a footprint, a hat, or a witness report—provides some information about who the culprit might be. You start with a *prior* suspicion about each suspect based on their history or other background information. This initial suspicion represents your *prior* probability, which reflects how likely you believe each suspect is guilty before considering any evidence.

As you uncover clues, you use Bayes' rule to update these suspicions. For each clue, you calculate how likely it is that a particular suspect would have left that piece of evidence, assuming they are guilty. Bayes' rule combines this likelihood with your prior probability to give a new probability, known as the **posterior** probability, for each suspect. The process repeats for each new clue you find.

The critical simplification in a *naive Bayes classifier* is the assumption that all the pieces of evidence are independent of each other, even if, in reality, they might be related. This assumption enables the multiplication of the likelihoods of each piece of evidence together for each suspect, updating their posterior probabilities step by step. After considering all the clues, the suspect with the highest posterior probability is the one that can be concluded is most likely to be the culprit.

Thus, the core of Bayesian learning involves starting with a *prior* distribution over the model parameters, representing our initial beliefs before seeing any data. As data is observed, this prior is updated to form a *posterior* distribution using Bayes' rule. This posterior distribution combines prior knowledge with data information, effectively balancing our initial beliefs with the observed evidence. The resulting posterior can then be used to make predictions on new, unseen data by integrating over all possible parameter values weighted by their posterior probabilities. This approach naturally handles uncertainty and allows for probabilistic predictions.

Despite their theoretical importance, the practical use of Bayesian methods in machine learning—pioneered by Judea Pearl—did npt begin in earnest until the latter half of the 20th century, with the development of naive Bayes classifiers and Bayesian decision theory in the 1960s and 1970s. [269]

The 1980s and early 1990s saw further development of Bayesian networks and the application of Markov chain Monte Carlo (MCMC) algorithms, which enabled sampling from posterior distributions in complex models. [130] The subsequent introduction of approximate inference methods, such as *variational inference* in the 1990s, made Bayesian approaches computationally feasible for larger data sets, further driving their adoption. [168]

The 2000s saw a further rise in the popularity of Bayesian methods with the advent of probabilistic graphical models [187] and hidden Markov models (HMMs). These approaches were widely used in fields like speech recognition, natural language processing, and computer vision, and Bayesian methods became established as a critical toolset for ML.

In the 2010s, as deep learning surged in popularity (see Part 12), researchers like Yarin Gal and Zoubin Ghahramani advanced the use of Bayesian principles in deep learning to quantify uncertainty in models, and Bayesian methods found renewed interest through Bayesian Neural Networkss and Bayesian deep learning. [109]

Additionally, scalable methods like Bayesian optimization have been used for tasks such as hyperparameter tuning and black box optimization, further integrating Bayesian approaches into modern ML workflows. [110]

Notes

1. Gini impurity measures how often a randomly selected element of a set would be incorrectly labelled if it were labelled randomly and independently according to the distribution of labels in the set.

2. Information gain, also known as Kullback–Leibler (KL) divergence, measures the amount of information gained about a random variable or signal from observing another random variable. In Decision Trees, the term information gain is sometimes used interchangeably with mutual information, although formally mutual information $I(X;Y)$ between two random variables X and Y is the Kullback–Leibler divergence of the joint distribution $p(x,y)$ from the product of their marginal distributions $p(x)p(y)$. [127]

3. Each particle in PSO represents a possible solution and moves around the solution space, searching for the best point, which will encode the solution to the optimization problem to be solved.

4. A tensor is a mathematical object that generalizes scalars, vectors, and matrices to higher dimensions. It is used to represent data with more than two dimensions and can be thought of as a multi-dimensional array.

5. An RBF network only ever needs to utilize **one** hidden layer.

6. These include: *spiking neurons* as widely used in computational neuroscience [145]; *kernel functions* as deployed in *radial basis function* networks [49], and *support vector machines* [43]; *gated MCP cells*, as deployed in LSTM networks [144]; *n-tuple* or *RAM* neurons, as used in *weightless neural network* architectures [38] and *stochastic diffusion processes* [30] as deployed in the *NESTOR multi-variate connectionist framework*. [256]

7. An active weight is a weight connected to a non-zero input.

8. Fundamentally, single-layer perceptrons can only learn to classify functions which are linearly separable; functions that are not linearly separable, famously include the XOR and parity functions of Boolean logic.

9. Although Hopfield received the Nobel Prize in physics for this contribution in 2024, Jürgen Schmidhuber points to earlier work by Shun-Ichi Amari in 1972 that reached similar conclusions. [8]

10. The PDP group's early work on Boltzmann machines is one of the earliest forms of generative AI, now ubiquitous in systems such as ChatGPT, Claude, Gemini or Deepseek, albeit the first seeds of generative AI can also be traced back to work from A. G. Ivakhnenko and V. G. Lapa (1965) and Amari (1972). [161] [8]

11. Although tracing the historical development of backpropagation is complex, and controversial (cf. Wikipedia https://tinyurl.com/2axhrrt4), the birth of modern backpropagation has also been positioned with Finnish mathematician Seppo Linnainmaa's reverse mode of automatic differentiation, published in 1970 as a method for efficiently calculating the derivative of a composite function by credit assignment in networks of differentiable nodes. [213]

12. A cell's *receptive field* refers to the specific region of sensory space (such as a part of the visual field, skin surface, or auditory range) where the presence of a stimulus influences the cell's behaviour.

13. A vivid description of the intellectual history surrounding Bayes, his discovery and its importance can be found in [323].

8 The Second and Third DARPA Grand Challenges

8.1 THE SECOND DARPA (DESERT) GRAND CHALLENGE

The second DARPA Grand Challenge took place 18 months after the first, once again in the Mojave Desert. The Red Team's goal was to learn from their previous mistakes and refine the existing strategy, with a focus on enhancing mapping, navigation, and hardware capabilities.

Sandstorm received a comprehensive makeover, featuring fewer sensors, simplified cable routing, a smaller and more movable gimbal for the lidars, and accelerometers to measure the movements of the chassis and dampers. William Red Whittaker and Chris Urmson also brought *H1ghlander*, a new Humvee equipped with basically the same hardware and software as its older sibling.

The mapping effort was greatly expanded. The Red Team deployed three drivers who spent an entire month scanning the desert for potential routes, covering 2,000 miles of terrain in great detail. As in previous years, a human team preprocessed the route just before the race to allow a computer to plan the path, which was then uploaded to the car. The team also introduced specific rules to prevent deadlock situations, such as the one that had occurred on the berm the previous year, which had so dramatically ended their race.

The improvements were successful. After qualifying rounds for the 2005 challenge, *H1ghlander* was in pole position, followed by *Stanley* and the Red Team's *Sandstorm.* Although both of the Red Team's cars completed the entire 132-mile (244-km) course in under seven hours, in the race, H1ghlander was overtaken by Stanley, a new vehicle from Stanford University in collaboration with Volkswagen's Electronics Research Laboratory (ERL), and it was this vehicle which eventually secured first place. Sandstorm finished second, approximately nine minutes ahead of H1ghlander. Whittaker later remarked:

> 'Sandstorm was a surprising second because its strategy was to creep for a reliable finish, and only the winner beat its pace ... Sandstorm was sent forth with a conservative plan, understanding that it would operate well below its peak abilities. But it was right on the leader's tail. It ran a perfect race.'

Stanley, designed by a team led by Sebastian Thrun, a German AI pioneer who would emerge as a key figure in the evolving AD industry, completed the course in 6 hours, 53 minutes, and 8 seconds, driving at an average speed of 19.1 mph (30.7km/h). Two other vehicles also finished the race, with a fifth completing it beyond the ten-hour time limit.

With five vehicles completing the course, Tony Tether achieved his first presentable success. The race results demonstrated that the gigantic eye approach triumphed

DOI: 10.1201/9781003561460-11

over the best map of the world, and neural networks emerged as a critical performance enabler. By focusing on machine learning, the Stanford team helped elevate AV performance to the next level.

Thrun's critical insight was that prioritizing mapping over vision was a misallocation of resources. As Alex Davies noted, Thrun had a straightforward diagnosis: The vehicles 'didn't see anything'. [77] Thrun realized that perfecting the vision system of the gigantic eye would eliminate the need for manual, error-prone labour. Even more crucial was Thrun's belief that machine learning would be the most effective solution to address the problem. As the Stanford DARPA team concluded in 2006:

> 'The road appearance is affected by a number of factors that are not easily measured and change over time, such as the surface material of the road, lighting conditions, dust on the lens of the camera, and so on ... This suggests that an adaptive approach is necessary, in which the image interpretation changes as the vehicle moves and conditions change.' [342]

Thrun's adaptive vision approach echoed Rodney Brooks's *active vision* insight—that it is easier to use the world as its own best representation and to sense the environment to get information from it as required, rather than to rely on the best internal representation—its map—of the world. Trying to anticipate and program every event in advance is incredibly burdensome and necessarily incomplete; this race showed that it is much better to train the agent to perceive and understand the environment as it unfolds.

The Stanford team tackled the problem by effectively starting from scratch and defining new ways to apply machine learning, statistical estimation, and probabilistic analysis to the vision problem. Sean Gerrish described the situation as follows:

> 'The Stanford team wasn't the first research group to use machine learning to design self-driving cars: Machine learning had been researched in the context of self-driving cars since at least the 1980s. But they were one of the first modern self-driving car teams to take such a full bet on machine learning, embracing its role in self-driving cars nearly a decade before it became a buzzword in the mainstream media.' [113]

In a summarizing paper published in 2006, Sebastian Thrun and colleagues assessed their approach:

> 'From a broad perspective, Stanley's software mirrors common methodology in AV control. However, many of the individual modules relied on state-of-the-art artificial intelligence techniques. The pervasive use of machine learning, both ahead and during the race, made Stanley robust and precise. We believe that those techniques, along with the extensive testing that took place, contributed significantly to Stanley's success in this race.' [342]

With hindsight, Thrun articulated this point even more concisely: 'Thanks to our emphasis on machine learning and adaptive intelligence, Stanley won.' [44] Stanley was based on a Volkswagen Touareg car, packed with processing power and multiple sensors. To be competitive, Stanley had to operate at a speed in excess of 25 mph. Thrun's overarching design principle was to treat autonomous navigation primarily as a software rather than a hardware problem. This approach prioritized hardware simplicity over functionality, leading, for example, to a new approach to laser sensor placement.

The Stanford architecture consisted of a hardware layer that handled both sensing [perception, localization] and actuation [the control of speed, braking, and steering]. A software planning module provided the necessary intelligent analysis and subsequent control instructions to the hardware layer.

The planning module responded to information about obstacles in the robot's path, evaluating options to avoid them and making adjustments to steering and speed, which were then sent to the actuator hardware controllers.

The Stanford team's solution centred on integrating—or *fusing*—data from multiple sensors to improve reliability. They combined visual information from cameras with more reliable distance measurements from lidar, using a Kalman filter to smooth out noise in the data. This adaptive vision approach was critical in improving the system's accuracy.

Stanley's navigation system relied on sensors, data processing techniques, and probabilistic models to ensure accurate navigation despite uncertain conditions and sensor noise. The obstacle detection system used five *roof-mounted* laser range finders to generate 3D point clouds, which were projected into a global coordinate frame.

The terrain classification process organized the 3D point data into a 2D grid map, classifying each cell as *drivable*, occupied by *obstacles*, or *unknown*. Stanley classified each grid cell by comparing the vertical distance between nearby points. If the height difference exceeded a critical threshold, the cell was marked as occupied by an obstacle. If no such difference was detected and at least one laser reading fell into the cell, it was marked as drivable. Cells without any laser data were marked as unknown. This 2D grid was continually updated and used by Stanley's navigation engine to plan a safe route.

A significant challenge in using laser data for terrain classification was managing errors in the vehicle's pose estimation—specifically, small roll or pitch errors. Because the lasers were scanning up to 30 meters ahead of the car, even minor pose inaccuracies could result in significant misclassifications of the terrain, leading to the detection of obstacles where none existed. Thrun deployed a probabilistic model to account for such pose estimation errors, helping Stanley to better differentiate between natural obstacles and false positives.

Stanley also incorporated machine learning techniques to fine-tune the system's parameters. A discriminative learning algorithm was employed to adjust parameters, including the critical height threshold for detecting obstacles. This supervised machine learning algorithm relied on a labelled training data set where the terrain was marked as *drivable* or *non-drivable* based on human driving patterns. Human drivers were enlisted to traverse various terrains, collecting lidar data as they went. The terrain that humans drove across was labelled as *safe*, while undriven terrain was considered *potentially hazardous*. This data was then used to fine-tune Stanley's obstacle-detection algorithms further.

The process involved a *coordinate ascent algorithm* that iteratively adjusted each parameter and evaluated its impact on classification performance. This tuning process enabled Stanley to achieve a significant improvement in terrain classification, reducing false positives from 12.6% to just 0.002% while maintaining a consistent rate of detecting natural obstacles.

Although the desert environment still had fewer transient complications than an urban environment, Thrun's new approach was state-of-the-art, bringing Stanley victory and the Stanford team two million dollars in bonus money.

Nonetheless, as sophisticated as the improved architectures were in 2005, they were still rudimentary compared to today's self-driving vehicles. Substantial problems remained to be overcome, as Sebastian Thrun and colleagues foregrounded:

> 'While the DARPA Grand Challenge was a milestone in the quest for self-driving cars, it left open a number of important problems. Most important among those was the fact that the race environment was static. Stanley is unable to navigate in traffic. For autonomous cars to succeed, robots, such as Stanley, must be able to perceive and interact with moving traffic.' [342]

This was to be the goal of DARPA's next and final challenge for AVs, the *Urban Challenge*.

8.2 THE THIRD DARPA (URBAN) GRAND CHALLENGE

In 2007, DARPA introduced a new challenge in urban environments with moving traffic. [249] Held on the streets of a decommissioned military base in California, autonomous vehicles were equipped with a sophisticated digital map of their environment, known as a road network definition file (RNDF). This RNDF was far more detailed than a typical GPS map, containing precise geometric information about lanes, lane markings, stop signs, parking lots, and designated checkpoints. To complement this digital data, teams also received high-resolution aerial imagery of the area in advance, which could be preprocessed, providing further data to refine and enhance their RNDFs prior to the event.

The third DARPA Grand Challenge, often called the *Urban Challenge*, required vehicles to navigate a 55-mile (97 km) course, and was considerably more complex than the two previous challenges—instead of a race against the clock, vehicles were tasked with completing multiple missions, each defined as a sequence of checkpoints. In the earlier competitions, vehicles started individually, and there were no other vehicles on the track with which the cars would have had to interact; this was the first time that multiple AVs had to operate jointly in the same local environment.

In the previous desert competitions, it was not essential to classify objects. Hence, vision systems process all objects in much the same way. The simulated city of the military base was very different. Besides traffic, there were pedestrians, road signs, traffic lights, road markings, etc. Driving on such streets requires *interpreting* the environment rather than simply *observing* it. In real urban mixed traffic, with pedestrians, cyclists, scooters, people in wheelchairs, and others, the environmental complexity is even higher, as unexpected cases may arise that can't be assigned to any existing object category. In the words of Sebastian Thrun of the Stanford team:

> 'In the last Grand Challenge, it didn't really matter whether an obstacle was a rock or a bush because either way, you'd just drive around it. The current challenge is to move from just sensing the environment to understanding the environment . . . you could claim that moving from pixelated perception, where the robot looks at sensor data, to understanding and predicting the environment is a holy grail of artificial intelligence.'

The complexity of the Urban Challenge was amplified by the fact that several robotic vehicles operated simultaneously in the same environment, sometimes under different speed restrictions. Competition vehicles were required to execute left turns, merge into high-speed oncoming traffic, and manage intersection precedence at multi-way stops. In addition, they needed to navigate around parked or slow-moving vehicles, park within designated lots, and perform U-turns when faced with complete road blockages.

Crucially, when encountering other vehicles, each robot was required to adhere to standard traffic *conventions and rules*, just like human drivers. To ensure safety and maintain control, vehicle speeds were generally capped at 30 mph, with even lower limits imposed in certain areas. This speed restriction allowed for thorough testing of the vehicles' capabilities without unnecessary risk.

The requirement for vehicles to correctly follow traffic rules and conventions added yet further complications. While there were no traffic lights on the site, there were several 'four-way' stops.[1] Here, the rules are clear for humans but hard for machines to apply appropriately in context.

The first vehicle to halt at a four-way stop always has the right of way. If two or three motorists reach a four-way stop simultaneously, each driver must yield to the driver to their right. That means the driver without a vehicle in the intersection to their right would have the right of way and go first. This rule remains in effect until all vehicles have safely passed through the intersection. However, if both vehicles are facing each other at the intersection and plan to proceed straight, they can pass through the intersection simultaneously.

Conversely, if two or more vehicles are all turning right, all vehicles can proceed with caution and pass through the intersection. Drivers may also be able to turn right when a car is going straight, provided they won't cross paths with the driver with the right of way. In these circumstances, drivers should be alert and check to see if other drivers are also entering the intersection to avoid collisions. When in doubt, drivers should wait until they have the right of way and use eye contact or signals to communicate as required, ensuring orderly and safe movement through the intersection.

But what happens if four cars show up simultaneously, each to the right of the other? This is a challenging problem for an AV to solve because, after it comes to a stop, the AV is unable to negotiate as a human driver would, e.g. wave or use the horn, and hence, is unable to move on. As Alex Davies reported, such a situation has actually occurred— *'four cars reached an intersection at the same time, leaving each in confusion about whose turn it was to go, creating history's first robot traffic jam'*—and required human intervention to resolve. [77]

Competition conditions thus demanded a wide range of specific manoeuvres from the autonomous vehicles. Each task represented a significant challenge in the realm of autonomous driving, requiring new hardware, sophisticated algorithms and improved decision-making processes, severely testing AV capabilities in real-world scenarios.

In the previous DARPA desert race, single-beam laser scanners were adequate for the challenge. However, in the new urban environment, a more detailed perception of the surroundings was critical. Velodyne, originally a manufacturer of high-end car speakers, recognized this shift early on. David Hall, the company's co-founder, anticipated that autonomous vehicles (AVs) would require a new type of sensor, prompting him to redirect the company's focus.

In the 2005 DARPA Grand Challenge, Hall introduced an early version of Velodyne's lidar system, mounted on a Toyota Tundra. Although the car failed to finish that race, its spinning lidar became the basis for Velodyne's HDL-64E model, which was unveiled for the 2007 Urban Challenge. This marked a significant breakthrough as this model featured a 360-degree field of view with 64 laser beams emitting infrared pulses, generating 1.3 million data points per second. The result was highly detailed 3D point clouds with a range of up to 120 meters, enabling precise object detection and environmental mapping.

Although widely used in the 2007 challenge, the Velodyne HDL-64E's high cost—*typically around $75,000* per unit—restricted its accessibility to competitors. Nevertheless, for those teams that could afford it, the HDL-64E provided vehicles with a comprehensive, high-resolution view of their surroundings from all directions.

In addition to helping vehicles perform tasks such as navigating complex urban environments, obstacle avoidance and interacting safely with other cars, the detailed 3D maps generated from its sensor data enabled vehicles to merge into traffic flows, behave appropriately at intersections, and avoid pedestrians and other road users. Indeed, the power of Velodyne lidar was fully evident in the 2007 challenge, where five of the twelve finishing vehicles utilized it. This solidified Velodyne's role as a key technology enabler in the future development of autonomous driving systems.

Due to the increased complexity of the 2007 Urban Challenge, DARPA held several qualification rounds before the race, admitting only eleven cars to the final event. Furthermore, the increased technical demands were reflected in the composition of the teams. The Red team of the first two challenges consisted of undergraduates, graduate students, and volunteers; now, there was a substantial team comprised entirely of senior researchers. Also, in this challenge, DARPA distinguished between professional teams and independent participants, who were treated differently and started separately. Thus, the third challenge demonstrated a substantial shift towards professionalization and industrialization.

In the 2007 challenge, the top teams from Stanford and Carnegie Mellon University faced off once again. Stanford's team, led by Sebastian Thrun, competed with their new vehicle, *Junior*, a reference to the university's full name, Leland Stanford *Junior* University. Like the Stanford vehicle deployed in the 2005 challenge, Stanley Junior was also a Volkswagen, but this time a Passat estate car.

Junior's new asynchronous pipelined architecture enabled its efficient operation in dynamic urban driving scenarios by processing sensor data and control commands concurrently. The system was designed as a distributed software pipeline, where different modules work independently and communicate through a publish/subscribe protocol. This allowed various processes, such as perception, planning, and control, to operate independently of one another, thereby reducing delays.

The architecture was divided into several key components. Sensor interfaces manage communication between the vehicle and its sensors, making the data available to other modules. *Perception* modules were responsible for segmenting the environment into static and moving objects and providing accurate localization relative to a digital map. *Navigation* modules determine the vehicle's behaviour through a combination of motion planners and a finite state machine, which aids in decision-making and prevents deadlocks. A custom drive-by-wire interface, developed by Volkswagen's Electronic Research Laboratory, managed the actual vehicle *controls*, such as steering, braking, and acceleration, converting software commands into physical actions. Finally, the *global services* module provides system-wide functions, including logging, time-stamping, and monitoring the health of the various modules. The innovations and strategies deployed in DARPA's third (urban) Grand Challenge have evolved into architectural principles that are still being used in today's robotaxis.

One of the critical features of Junior's architecture is its low latency, with sensor data being processed and acted upon within approximately 300 milliseconds. This is crucial for responding to real-time driving conditions. By organizing the software into asynchronous modules, each continuously processing data, the vehicle can handle complex driving tasks such as obstacle avoidance, lane changes, and merging into traffic. This approach allowed Junior to perform reliably even in the presence of other vehicles and changing environmental conditions. The use of asynchronous processing ensures that the system remains responsive, flexible, and capable of handling the diverse challenges of urban navigation.

In contrast, Chris Urmson, now a professor at CMU, led his university's team, rebranded as *Tartan Racing*. The team was composed of students and staff working alongside researchers from General Motors, Caterpillar, Continental, and Intel. In the previous challenges, Urmson and Whitaker deployed a military Humvee called *H1ghlander*, utilizing their 'best map of the world' approach. However, it was clear that in the 2007 challenge, this would no longer suffice due to the additional driving complexities that arise when interacting with other vehicles on the road. So the Humvee was replaced by a Chevrolet Tahoe named Boss in honour of Charles Boss Kettering, the renowned GM researcher—and also likely a nod to their leadership ambitions.

In addition, with a high level of integration between high-level planning and low-level control, as explored in Eran Gat's ATLANTIS architecture, the CMU team elected to utilize a sophisticated three-layer modular architecture for navigation and control. This comprised a *mission planning* layer, a *behavioural reasoning* layer, and a *motion planning* layer. Together, these layers allowed Boss to drive safely through the new, more complex urban scenarios, including interacting with traffic, obeying road rules, and avoiding obstacles, all without human intervention.

The mission-planning layer was responsible for determining the overall route to mission checkpoints. It worked similarly to how a human driver plans a route, calculating the best path based on road conditions, speed limits, and potential blockages.

Boss used a predefined road network derived from the road network definition file to create a graph of possible routes. Each waypoint in the network became a node in the graph, and connections between these nodes represented possible lanes. The mission planner continuously updated the graph based on real-time sensor data, such as road closures or traffic, allowing Boss to re-plan and find alternative routes as necessary. If all known routes were blocked, the car could revisit previously blocked paths to see if any obstacles had been cleared.

The behavioural reasoning layer translated the strategic goals from the mission planner into tactical decisions. This layer managed lane changes, intersection handling, and navigation through unstructured environments, such as car parks. It operated within three primary contexts: *Road driving*, *intersection handling*, and *zone navigation*.

Road driving involves tasks such as changing lanes, maintaining a safe distance from other vehicles, and responding to changing traffic conditions. Intersection handling ensured that Boss adhered to traffic laws, such as stopping at intersections and giving way to other cars. In unstructured environments, such as parking lots or when manoeuvring through tight spaces, the zone navigation component took control. The behavioural reasoning system was also crucial for handling error recovery and adapting to unexpected events, such as blocked roads or unforeseen obstacles, by selecting alternative strategies.

The motion planning layer executed the tactical decisions made by the behavioural layer. It generated multiple potential trajectories and evaluated each one based on safety, dynamic feasibility, and efficiency. The system employed a model-predictive approach, predicting how the vehicle would respond to various control inputs and selecting the optimal trajectory based on the current environment.

For structured road driving, the motion planner ensures the vehicle stays within its lane while avoiding other cars and obstacles. In unstructured environments, such as parking lots, the planner employed a lattice-based search, taking into account the vehicle's position, orientation, and speed to determine a safe path through the space. The planner continuously adjusted the vehicle's trajectory in real-time as new sensor data became available, ensuring it could adapt to dynamic changes in the environment.

Each layer of Boss's architecture was closely coupled with the others. The mission planner sets strategic goals, the behavioural reasoning layer translates these into tactical actions, and the motion planner executes these actions while continuously updating the other layers on progress and any encountered obstacles. Thanks to this layered architecture, Boss's ability to handle complex driving tasks in real time enabled it to navigate urban environments autonomously and complete the 2007 DARPA (urban) Grand Challenge.

As we have seen, Boss and Junior approached navigation and driving in urban environments with very distinct navigation and control strategies. Boss utilized a three-layered architecture for its control system, which included mission planning for long-term route selection, behavioural planning for tactical decisions, and motion planning to generate precise trajectories. Junior, on the other hand, employed a data-driven pipeline where perception, planning, and control were handled by separate modules that communicated asynchronously. This modular design allowed for a more flexible and distributed decision-making process, contrasting with Boss's more hierarchical approach.

In terms of perception and sensor fusion, both vehicles relied heavily on lidar and radar systems, but they differed in their implementations. While both teams used the Velodyne 64-beam lidar to obtain a highly detailed 3D view of the environment, Boss used a combination of lidar, radar, and cameras to detect and track obstacles, integrating this information into a cohesive model of its surroundings. In contrast, Junior employed multiple lasers and radars, placing a strong emphasis on grid-based mapping and real-time terrain analysis, which made it particularly adept at identifying fine features, such as curbs and berms. This gave Junior an edge in precise localization and handling complex road environments.

Obstacle detection and tracking also revealed differences in approach. Boss relied on its sensor data to make predictions about the movement of other vehicles and objects, using a *Simplified Bicycle Model*—that reduced cars to a simple 'two wheeled model' for more efficient real-time simulation—alongside an extended Kalman filter to accurately forecast the motion of dynamic obstacles.[2] [272] Junior, by contrast, used a *particle filter algorithm* to generate *multiple hypotheses* about the movement of dynamic objects, focusing on the most likely outcomes.[3] The latter method was more adaptable in highly dynamic environments, enabling Junior to track and predict the motion of surrounding vehicles with great flexibility.

When it came to decision-making, Boss's behavioural system integrated tightly with its motion planning, ensuring smooth transitions between high-level decisions and low-level trajectory execution. This allowed Boss to make tactical decisions, such as when to overtake or yield at intersections, fluidly. Junior, however, relied on a finite state machine to manage its behaviour, with each state corresponding to a specific driving task, such as handling intersections, lane changes, or parking. The FSM also included recovery behaviours, allowing Junior to escape from deadlocks or navigate complex situations, such as blocked roads, using a hybrid A* planner for intricate manoeuvres like U-turns and parking.

Both vehicles demonstrated strong capabilities in handling road blockages and unstructured environments, but their approaches differed. Boss emphasized error recovery, using the replanning systems to navigate around obstacles or blocked roads by executing U-turns or rerouting as necessary. In contrast, Junior, with its hybrid A* planner, was highly effective when operating in unstructured environments, such as parking lots, where predefined lanes were absent. The planner enabled Junior to generate arbitrary but realisable trajectories, providing it with superior manoeuvrability in unstructured, less-constrained environments.

At intersections and during lane navigation, Boss was able to localize itself with high precision relative to its high-quality, preloaded map, making it very effective at maintaining lane discipline and obeying traffic rules. Junior, however, continuously refined its position relative to the road network definition file through a lateral offset correction mechanism. This allowed Junior to dynamically adjust its position based on real-time sensor data, which was particularly helpful when road conditions deviated from its preloaded map. Junior's system for managing intersections also involved critical zone monitoring, where it would wait for specific zones to clear before proceeding, adding an extra layer of safety in high-traffic areas.

In terms of free-form navigation, Junior's hybrid A* planner gave it an advantage in handling complex driving tasks such as manoeuvring in a car park or navigating through blocked roads. The planner allowed Junior to explore different paths in real-time, ensuring that it could adapt to changing environments more effectively than Boss's structured motion planning system. This flexibility made Junior better suited to environments where road layouts were unclear or where multiple obstacles were present.

Lastly, error recovery was a critical aspect of both systems; however, Junior's approach was more comprehensive. Boss could handle errors by replanning routes or executing U-turns, but Junior's FSM-based system incorporated more detailed recovery behaviours. If Junior detected that it was stuck, it could invoke the hybrid A* planner to find alternative paths, and in extreme cases, it could even skip checkpoints to avoid infinite loops. This made Junior highly resilient in complex environments, capable of recovering from a wide range of unexpected situations, including dealing with inaccurate road descriptions or blocked passages.

Nevertheless, by efficiently deploying their optimized tripartite model approach—*a static obstacle map, a list of the moving vehicles in the world, and the location of Boss relative to the road*—CMU was able to claim the throne back. Boss won with an average speed of 14 mph. Junior, the car from the Stanford team, ranked second. Its average speed while driving was 13.7 mph.

However, it was not just Boss and Junior who completed DARPA's 2007 (urban) Grand Challenge; four other vehicles crossed the finish line: Six out of eleven cars completed the course, a result that exceeded the most optimistic contemporary expectations.

In retrospect, several factors proved decisive for the breakthrough. A change of emphasis effectively formulated all control requirements as software problems and exploited improved digital technology, such as significantly cheaper and larger computer memory. On-the-fly perception, achieved by fusing data from combinations of sensors, enabled AVs to move more safely in a dynamic environment. The key breakthrough, however, was the application of machine learning in conjunction with innovative approaches to mapping dynamic environments.

The success of machine learning becomes more apparent when contrasted with rule-based approaches from the same race. This comparison yields two more subtle lessons. The more obvious one: *In an open environment, handcrafting rules for all contexts is all but impossible.* The other, less obvious, but even more profound principle is: *Never try to accomplish more in system design than is required.* The latter entailed focusing on engineering appropriate narrow AIs, not pipe dreams of AGI.

Meredith Broussard, a reporter covering the 2007 Urban Challenge and an associate professor of data journalism in New York, focused on *Little Ben*, [39] the fourth-placed car, which finished outside the six-hour race time limit, to illustrate this point. [50] The Franklin Racing Team designed Little Ben to deploy a form of knowledge-based AI—a GOFAI approach. Their goal was to create a machine that could make flexible decisions based on its understanding of the world by using fixed, pre-programmed rules and a sizable handcrafted knowledge repository. It was designed to map a person's intelligence onto appropriate symbolic representations to develop a generalized, rule-based AI that could cope seamlessly and intelligently with any new situation.

In order to achieve this, Little Ben attempted to *see* objects in a similar way to humans. It was, therefore, necessary to imbue the car with some form of *theory of mind* in order to be able to reason about the intentions of other entities.

A lidar system mounted on the roof captured data that its software would later use to identify the object the data corresponded to, e.g. by analysing its shape, colour, size, etc., before deciding what it should do next. For example, if it recognized a person or dog, it would slow down and/or modify its planned path. Conversely, if it identified a bird, it would assume the bird would fly away well before any collision and continue without deviation from its planned trajectory.

To deliver the above, Little Ben needed extensive information about objects in the real world, including their shapes and behaviours. However, this approach had a fundamental problem: In practice, handcrafted rules cannot anticipate the potential infinitude of exceptions to general principles.

Consider what could happen if, say, a random object, such as a traffic cone, were knocked over. As Little Ben was unable to rotate unknown objects in three dimensions, it would no longer be able to identify the object as a traffic cone and hence be unable to assign its likely behaviour correctly.

In contrast to Little Ben, which had to identify all objects in sight—the road, birds, pedestrians, buildings, traffic cones, etc., *and run predictions for each scenario*—the CMU and Stanford researchers chose to circumvent this issue altogether. Drawing inspiration from the Wright brothers' pioneering insight that an aeroplane does not need to flap wings to fly, they concluded that a car would neither need human-like sentience nor complex computational ontologies to drive.[4]

Instead, the approach involved pre-loading vehicles with detailed maps annotated with comprehensive descriptions of all static obstacles and predefined routes. AI and ML algorithms were deployed to process data from lidar, radar, and cameras to generate a *dynamic map* of transient objects. Each object was characterized by a mathematical model predicting its likely movement. Efficient algorithms were employed to prioritize objects relevant to driving trajectories, ensuring computational focus on elements critical to navigation.

The machine-learning approach marked the tremendous progress that the field had seen across the three DARPA competitions. The critical success factors included software-focused approaches, the availability of inexpensive and large computer memory, real-time perception through sensing, the application of machine learning and probability theory, and a strong focus on narrow AI approaches. With these innovations, AV had catapulted from failure in desert scenarios to mastering inner-city mixed traffic in a relatively short time. And so it appeared that if technological developments continued at this pace, the realization of fully autonomous mobility would only be a matter of will. At last, Geddes' vision of completely driverless inner cities seemed almost within reach.

Notes

1. Common in America, a four-way stop is an intersection where all approaching vehicles **must** stop, typically controlled by stop signs on all approaches.

2. Instead of modelling all four wheels, the simplified bicycle model reduces the car to two wheels—one at the front for steering and one at the rear. The model utilizes the vehicle's speed, steering angle, and orientation to predict its movement in a 2D space. It is helpful in estimating the future position of a vehicle in real-time, such as in autonomous driving systems, because it simplifies complex vehicle dynamics into manageable equations, making calculations faster and more efficient.

3. A Particle Filter Algorithm is used to estimate the state of a system over time by representing possible states as particles. [120] Each particle predicts a potential state, and as new observations are made, the algorithm updates these particles by assigning weights based on how well they match the observations. Particles with higher weights are more likely to represent the actual state, and the algorithm resamples particles, focusing on the most likely outcomes. This process allows the filter to refine its estimate of the system's state, making it practical for tracking in uncertain environments.

4. A computational ontology is a structured framework that defines the relationships between concepts in a particular domain, allowing computers to understand and process information more effectively.

Section IV

Statistics, Big Data, and Deep Learning

As Meredith Broussard eloquently described, today's self-driving cars use a *statistical approach* to process sensor data and make decisions about how to navigate their environment. [50] They do not need to *understand* the rules of driving the way a human driver does, but instead rely on deep learning algorithms that are trained on vast data sets. This allows the car to recognize patterns and make predictions based on probabilities, instead of a large set of human-engineered *rules* for driving.

Deep neural networks (DNNs) define a subset of machine-learning models that extend the concept of the multilayer perceptron.[1] By incorporating many more hidden layers between the input and output, they become *deep* networks, hence the name. These additional layers enable DNNs to learn and model complex, high-dimensional data.

In contrast to conventional feedforward multilayer perceptrons, which typically consist of an input layer, one or two hidden layers, and an output layer, a basic feedforward DNN can have dozens or even hundreds of hidden layers. Each layer functions to extract increasingly *abstract* features from the data, enabling the network to represent and process highly intricate patterns and relationships.

This hierarchical approach to feature abstraction has been demonstrated to be particularly useful for tasks such as image and speech recognition, natural language processing, and self-driving cars. Examples of feedforward DNNs include *deep multilayer perceptrons* and *deep convolutional networks*.

However, in addition to classical regression and discriminative ML architectures, the development of DNNs has led to new types of generative AI architectures (Gen-AI), which can be used to generate new random data instances of either (i) an observation x and target y, (x, y) or (ii) an observation x given a target y, $(y \rightarrow x)$. Because of their ability to create new data instances, Gen-AI architectures can be trained to generate outputs that are uncannily similar to real-world exemplars.

Examples of such architectures range from Geoff Hinton's early use of a classical Boltzmann machine to model probability distributions over binary random variables to image-from-text generators such as DALL-E and interactive chatbots like ChatGPT. Furthermore, given sufficient capacity, data, and training time, modern generative architectures can learn *any* distribution of data, hence the enormous variety of their potential use cases.

9 Theory, Empiricism, and Data

As data-driven AI becomes ever more ubiquitous, we appear to be moving into a world where we no longer need models or theories to explain how things work. This is the world that is envisioned by proponents of *Big Data*. In an influential article in the technology magazine Wired, titled *The End of Theory*, Chris Anderson, then-editor, argued against the need to discover models and causal processes, making a controversial assertion that, with enough data, correlation is sufficient to explain complex phenomena. Forget about theories of human behaviour and psychology; instead, the focus should be on finding patterns in data at the petabyte scale. As Anderson suggests:

> 'Information is not a matter of simple three- and four-dimensional taxonomy and order, but of dimensionally agnostic statistics. It calls for an entirely different approach, one that requires us to lose the tether of data as something that can be visualized in its totality. It forces us to view data mathematically first and establish a context for it later. [...] With enough data, the numbers speak for themselves.' [10]

But his idea is not new. As Francis Bacon, the British Empiricist and father of the scientific method, argued in the *Novum Organum*, knowledge should be based on experimental evidence rather than preconceived notions. [17] Bacon argued that deductive reasoning is ultimately limited because setting a premise ahead of an experiment constrains reasoning to suit that premise. His bottom-up approach was to appeal to empiricism: observations and data, rather than the deductive reasoning which had dominated science since Aristotle.

Similarly, as Anderson also observed, advocates of Big Data assert that data should be allowed to *speak for itself* without the need for models or hypotheses. [228] A very recent example—that Anderson could not have known at the time but which he would have prominently claimed for his view—is protein structure prediction. DeepMind's AlphaFold is able to predict the structure of complexes created by proteins with DNA, RNA, various ligands, and ions by letting an attention network statistically correlate data from 170,000 protein examples. This breakthrough earned its creators, represented by Demis Hassabis and John Jumper, the 2024 Nobel prize in chemistry.

This, and earlier examples, demonstrate how advances in computing power and the internet have enabled the gathering and analysis of vast amounts of data, ushering in the so-called *petabyte age*, where data can be processed and analysed at an unprecedented scale. At this scale, data becomes more than just a matter of simple categorization; it requires a new approach that is mathematically driven. The focus

DOI: 10.1201/9781003561460-13

shifts from searching for the logical formulae of causation to observing *correlations* or the relationships between data that cannot be explained by mere chance alone.

Such *model-free* access to knowledge alters the scientific process: In place of sampling strategies that focus on small, random parts of a problem, a comprehensive evaluation of the data as a whole enables the generation of a more complete picture. In turn, this increases tolerance for imprecision.

Now, rather than seeking accuracy under controlled and model-simplified conditions, the messiness in the data can be understood as a reflection of the complexity of nature. Measurement errors thus become less significant, and the focus shifts from searching for causalities to observing correlations. As Anderson concludes:

> 'Petabytes allow us to say: Correlation is enough. We can stop looking for models. We can analyse the data without hypotheses about what it might show. We can throw the numbers into the biggest computing clusters the world has ever seen and let statistical algorithms find patterns where science cannot. The new availability of huge amounts of data, along with the statistical tools to crunch these numbers, offers a whole new way of understanding the world. Correlation supersedes causation, and science can advance even without coherent models, unified theories, or really any mechanistic explanation at all.' [10]

In short, the era of Big Data represents a new way of understanding the world, one that is driven by vast amounts of data and mathematical algorithms. With this approach, we can uncover new insights and patterns that we previously did not know existed, without the need for explicit models or theories. This aspect of Big Data is hugely exciting and potentially promises to yield significant steps forward in the development of AI. This is the guiding philosophy behind Elon Musk's current thinking. Musk is undoubtedly a firm advocate of the *scaling is enough* approach as, although he has never made it explicit, his entire approach is geared towards maximizing data use and automating pattern recognition.

9.1 MASSIVELY PARALLEL PROCESSING

Computers have evolved significantly since the early days of computing, and modern AI systems would not exist without the necessary concomitant hardware. Sara Hooker from Google Research argues that a particular approach often succeeds because it aligns with available software and hardware rather than because the idea itself is superior. She refers to this phenomenon as the 'hardware lottery': the choices made about software and hardware have frequently determined the winners and losers in the history of computer science. [149]

This is particularly evident in the case of so-called deep neural networks. While their algorithmic foundations date back to the 1960s, it took another four decades for the hardware to catch up, highlighting the gap between theoretical advancements and practical computational capabilities.

A crucial factor in the evolution of computation hardware has been the development of new computer architectures and massively parallel processing. Richard

Sutton termed this the 'bitter lesson' of 70 years of AI research: AI's success is fundamentally tied to the provision of increasingly massive computing capacity rather than methods based on human knowledge and expertise. Methods that most effectively harness exponentially growing computation—search and learning—will consistently surpass human-centric approaches due to the exponentially scaling nature of Moore's law. [337]

One of the first, the so-called 'von Neumann architecture', was revolutionary when it was published in 1945. Its computing architecture consisted of a central processing unit (CPU), a relatively slow-to-access main storage area, and a relatively fast-access memory called random-access memory (RAM). The stored-program concept that it employed, in which instruction data and program data were both stored in the same memory, was both cost-effective and easy to implement.

However, the von Neumann architecture had a significant downside that became more pressing with increased data flow: As the processor and memory were separated, all data had to be transmitted between the two, leading to a frequent, energy-hungry data movement that was termed the von Neumann bottleneck.

Every piece of data and instruction has to pass across the data bus in order to move from the main memory into the CPU (and back again). This is a problem because, typically, the data bus operates at a substantially slower rate than the CPU can execute instructions. Hence, the von Neumann bottleneck. If no mitigating steps are taken, the CPU will spend a significant amount of time waiting. [324]

One mitigation strategy is to use a *cache*, effectively a holding area on the CPU side of the memory bus. Thus, an instruction cache would store the most frequently used program code in the holding area, eliminating the need to cross the bridge every time; similarly, a data cache would store the most commonly used data.

If the programmer is ingenious enough, they can make it easier for the CPU to store the most often used parts of the code in the program cache, an activity termed *code optimization*. Consider the pseudo-code below:

REPEAT; ADD TEN TO X; UNTIL FINISHED.

In this example, if the instruction ADD TEN TO X was stored in the program cache, it would not have to be fetched from memory each time, and the resulting code would run much faster.

Another method to mitigate the von Neumann bottleneck is to use a reduced instruction set computer (RISC) architecture. In addition to simplifying the machine-code instructions available to the programmer or compiler, RISC aims to reduce memory references by utilizing a relatively large number of on-chip registers. This design minimizes the need to access the main memory when running code.

In the early 1980s, Japan launched the *fifth-generation programme*, a national initiative to develop a massively parallel computer, as another way to overcome the von Neumann bottleneck. Although the program did not ultimately succeed, it was taken seriously by leading US scientists and accelerated research efforts at American institutions.

Parallel processing involves the use of multiple processors to perform coordinated computations simultaneously. This approach has gained prominence with the development and widespread use of graphics processing units (GPUs), parallel architectures designed to handle specialized tasks far more efficiently than conventional processors.

Initially developed to facilitate fast graphics rendering in computer games, GPUs were subsequently repurposed for machine learning, as the vector arithmetic central to game graphics is also fundamental to neural computing. In their specialist domain, GPUs are significantly more powerful than general-purpose CPUs, contributing to the recent massive valuation of chip manufacturers, first and foremost, their pioneer, Nvidia.

One of the first influential papers describing the application of GPUs to large neural networks was 'Large-scale Deep Unsupervised Learning Using Graphics Processors' by Rajat Raina, Anand Madhavan and Andrew Ng in 2009. [279] In 2010, Dan Ciresan achieved a breakthrough success with a GPU-based network on the MNIST database, [67] which AlexNet followed in the ImageNet competition in 2012. [190] The success of AlexNet in this competition rendered GPUs the de facto standard for training neural networks.

The Google Brain project, which in 2011 trained neural networks to recognize cats simply by processing a vast lake of visual data (scraped from movies on YouTube), required 2,000 CPUs; however, as Nvidia's Bryan Catanzaro and Andrew Ng demonstrated, only 12 GPUs would be needed to deliver the same level of performance. [155]

Indeed, such results led Yann LeCun, Yoshua Bengio and Geoffrey Hinton—sometimes apostrophized as the *godfathers of deep learning* because of a joint paper in Nature in 2015—to specifically note that the deep learning revolution was only made viable by the advent of *fast graphics processing units* (GPUs) that were convenient to program and allowed researchers to train networks using *large, easily available datasets* ten to twenty times faster than before.[2] [200]

9.2 THE UNREASONABLE EFFECTIVENESS OF DATA

The power of data and mathematics in AI is almost *inexplicable*, as physicist Eugene Wigner wryly observed at the beginning of his essay, *The Unreasonable Effectiveness of Mathematics in the Natural Sciences*: 'The world around us is of baffling complexity, and the most obvious fact about it is that we cannot predict the future.' [379] Yet, somehow, mathematical theories have predictive power in the description of physical laws of nature:

> '...it is important to point out that the mathematical formulation of the physicist's often crude experience leads in an uncanny number of cases to an amazingly accurate description of a large class of phenomena. This shows that the mathematical language has more to commend it than being the only language which we can speak; it shows that it is, in a very real sense, the correct language.' [379]

In contrast, Google's Alon Halevy, Peter Norvig, and Fernando Pereira foreground that this *unreasonable* effectiveness of mathematics seems to fail when confronted with the rich informality of human behaviour:

> 'Meanwhile, sciences that involve human beings rather than elementary particles have proven more resistant to elegant mathematics. Economists suffer from physics envy over their inability to neatly model human behaviour. An informal, incomplete grammar of the English language [278] runs over 1,700 pages.' [126]

...and instead advocate a return to *Empiricism*,[3] building the foundations of this move on what they, in an ironic nod to Wigner, term 'The Unreasonable Effectiveness of Data':

> 'Perhaps when it comes to natural language processing (NLP) and related fields, we're doomed to complex theories that will never have the elegance of physics equations. But if that's so, we should stop acting as if our goal is to author extremely elegant theories and instead embrace complexity and make use of the best ally we have: The unreasonable effectiveness of data.'

In other words, although theoretically, the number of grammatical English sentences is infinite, humans only ever derive a finite number of distinctions when using them. Hence, Halevy et al.'s intuition was that leveraging probabilities *learnt* from vast text corpora would be more effective in NLP than deploying a [relatively] small number of general linguistic *rules*; furthermore, noting that in executing this process: The first lesson of web-scale learning is to use *available* large-scale data rather than hoping for *annotated* data that isn't available (our emphasis).

Similarly, although not explicitly stated, it is evident that Elon Musk also applies empirical logic to the field of computer vision. Autonomous driving now relies heavily on Big Data and the remarkable effectiveness of statistical methods. Thus, rather than following predetermined driving rules, self-driving cars transform every object and trajectory in their environment into numerical data. By estimating probabilities and likelihoods from this data, they generate complex maps of obstacles and potential paths towards their goal.

Thus, in contrast to the linguistic turn that dominated Western analytic philosophy in the latter half of the twentieth century,[4] empiricism and the statistical turn now dominate AI research. Data has proved to be much more powerful than anyone imagined, as connectionists and other researchers push the limits of Big Data and statistical approaches to artificial intelligence.[5]

At the heart of the statistical turn lies statistics, and herein, this term defines a different meaning from probability. As Prof Steven Skiena clarifies, in the world of mathematics, two main branches deal with analysing data: probability and statistics. [325] Probability is concerned with predicting the likelihood of future events—*given the conditions, what is the likelihood of a certain outcome?*—while statistics deals with the analysis of past events—*given the data, what can we infer about the underlying population?*[6]

Thus, while probability is concerned with predicting the outcomes of processes in an *idealized or hypothetical world*, where the rules and parameters are known precisely, statistical theory, on the other hand, is used to analyse and interpret historical data from the *real world*—where the available information is often imperfect, noisy, or uncertain—to find patterns that can be mapped into the future. It allows us to infer the system's underlying properties and measure how closely our real-world observations align with those in the idealized model. It is concerned with real-world observations and emphasizes empiricism over theory. In summary, while probability theory enables us to find the future behaviour of a given *ideal* world, statistical theory allows us to measure the extent to which our world is *ideal*.

Language offers an excellent example of the methodological shift to Empiricism. In the early approaches to machine translation, for example, researchers tried to create formal models that defined all rules for grammar, syntax, and even semantics. These models heavily relied on propositional logic and linguistic insights. Researchers tried to use linguistic rules to build systems that *understood* how language works. However, as language is far too subtle and complex to be fully captured by a simple rule-based instruction, they failed.[7]

Progress in the field of NLP was slow until it moved away from models, logic, and rules. If you have enough data and the capacity to process it, it is now possible to translate language reasonably well. A ramification of this was trenchantly pinpointed in a sentence by Frederick Jelinek, an expert in NLP at IBM Research, who is quoted in the mid-80s as saying: 'Every time I fire a linguist, the performance of the speech recognizer goes up.' Although this citation was sharpened in later repetition, it very nicely illustrates a point: Up to today, statistical models have proved much more effective in capturing [linguistic] aspects of intelligence. [164]

Since the statistical turn in linguistics, many long-standing problems have been effectively resolved. The critical breakthrough lies in the concept of the word vector or word embedding, which forms the foundation of modern transformer neural architectures, such as those used by ChatGPT. This approach encodes each word, or token, as an n-dimensional real-valued vector, enabling, after Foucault, a *mathesis* of word meanings. [106]

While traditional approaches in natural language processing and AI in general relied on deterministic methods, modern NLP has adopted a more statistical perspective. This shift in strategy in linguistics precisely echoes that seen in computer vision following the success of deep learning in the ImageNet benchmark in 2012.

9.3 THE 2012 IMAGENET CHALLENGE

How do we teach machines to recognize mere patterns of pixels and features as objects in an image? Computer vision is one of the most extensively researched subfields of artificial intelligence. However, as Nils John Nilsson, Stanford's Kumagai Professor of Engineering and the creator of Shakey the robot, observed, early vision systems, such as those deployed in self-driving cars, were susceptible to environmental conditions. Factors such as lighting, the sun's position, and shadows significantly affected their performance, making it nearly impossible to anticipate every scenario in a rule-based system. These days, this is where machine learning comes into play.

To train machines to see using supervised learning, we need access to large amounts of labelled data from sample images. The most important public source for this data is the internet. Unfortunately, most of the pictures on the internet are *not* labelled. To solve this problem, Stanford professor Li Fei-Fei—who later moved to Google to lead its *Cloud AI* division—and her team decided to categorize internet images into carefully labelled domains. They used Google's *Image Search* to locate as many images as possible for each category, resulting in a collection of around 9,000 images that researchers could use to train and evaluate their algorithms.

The success of this project led to an even more ambitious one: *ImageNet.* Again, using Google's Image Search, ImageNet rapidly grew to over 14 million high-resolution images labelled with over 22,000 categories. This data set formed the basis of the 'ImageNet Large-Scale Visual Recognition Challenge', organized by the Stanford team in 2010. The challenge included 1.4 million images from 1,000 categories, and participants achieved minor, incremental improvements by deploying more and more handcrafted features, reaping low-hanging fruit. Hence, the error rate dropped from 28% in 2010 to 26% in 2011.

However, the pivotal breakthrough occurred in 2012 with advancements in machine learning, which revolutionized computer vision and were crucial for making self-driving cars viable. This breakthrough was closely tied to the 2012 ImageNet challenge, which was to showcase the power of deep learning in image recognition and classification tasks.

Notes

1. For a historical perspective, see Jürgen Schmidhuber (2022). [308, 309]

2. The possibility of dramatically optimizing the implementation of ANNs using GPUs was already emphasized by Keechul Jung and Kyoungsu Oh in 2004, well before the article in Nature. [266] In the paper, they describe accelerating traditional ANNs by a factor of twenty.

3. A philosophical theory that emphasizes the role of experience and sensory observation as the primary source of knowledge.

4. The *linguistic turn* in philosophy refers to a significant shift in the 20th century that emphasized the importance of language in understanding philosophical problems. This movement suggests that many traditional philosophical issues, particularly those related to meaning, knowledge, and reality, are deeply intertwined with how language functions. Instead of focusing on abstract metaphysical concepts, philosophers began analysing the structure and usage of language to clarify and resolve these issues.

5. It is essential to note the distinction between the mathematical approaches to intelligence, vis-à-vis probability and statistics and the *probabilistic school* of AI. While the former mathematically underpins the paradigm-shifting power of modern connectionist AI (as exemplified in deep learning, DCNs, etc.), the latter defines a set of radically different, non-connectionist, machine learning approaches, encompassing Bayesian models, probabilistic graphical models, hidden Markov chains, etc. And while both approaches continue to evolve, currently, AI, driven by deep learning and Big Data, is fast becoming ubiquitous.

6. To illustrate the difference between these two fields, Skiena, [325] opens up a hypothetical scenario involving a mathematician encountering a craps game for the first time. If the mathematician is a probabilist, they will approach the game by examining the dice and assuming that each face is equally likely to land

face up. Based on this assumption, they can then calculate the probability of winning or losing the game. However, if a statistician were to observe the same game, they would be more sceptical and wonder if the dice are truly fair or if they might be loaded. The statistician would observe the game over some time and keep track of how often each number comes up. This information would then be used to determine if the observations are consistent with the assumption of equal probability faces. Once the statistician is confident that the dice are fair, they may seek the advice of a probabilist to determine the best strategy for playing and winning the game.

7. This victory of Empiricism is by no means universally accepted within computational statistics, where scientists such as Walid Saba have trenchantly presented limits to empirical approaches that have roots in Fodor and Pylyshyn's early criticisms of Connectionism. [289]

10 New Forms of Learning

Machine learning has traditionally relied on supervised learning, where models are trained on labelled data sets. However, acquiring labelled data can be expensive, time-consuming, or impractical in many real-world applications, such as natural language processing, image recognition, and bioinformatics. On the other hand, unsupervised learning, which aims to uncover patterns and structures in unlabelled data, often struggles to achieve the accuracy and applicability of supervised approaches in predictive tasks. This limitation has driven researchers to explore methods that leverage both labelled and unlabelled data, leading to the emergence of new forms of *semi-supervised* and *self-supervised* learning.

Semi-supervised learning and self-supervised learning are both methods that focus on improving the performance of machine learning models by leveraging large corpora of readily available unlabelled data. By incorporating unlabelled data into the training process, these approaches aim to reduce reliance on labelled data sets and improve models' generalization capabilities. In this way, both have proven instrumental in fields such as computer vision, natural language processing, and speech recognition, where data is abundant but labelling is a bottleneck.

However, despite their shared goals, the two approaches differ significantly. Semi-supervised learning combines a small amount of labelled data with a large amount of unlabelled data to enhance task-specific performance, such as classification or regression directly. It requires *some labelled data* to guide the learning process and typically trains a single model that uses both labelled and unlabelled data simultaneously. Methods in semi-supervised learning often include label propagation, graph-based approaches, and generative models, such as semi-supervised generative adversarial networks (GANs).

In contrast, self-supervised learning is *entirely independent of labelled data* in its initial phase. It involves creating pseudo-labels or synthetic tasks (known as pretext tasks) from unlabelled data, enabling the model to learn general-purpose representations. These representations can then be fine-tuned for specific downstream tasks using a smaller labelled data set. Self-supervised learning often relies on techniques such as contrastive learning, masked autoencoders, or predictive tasks for problems like predicting missing parts of an image or reconstructing masked words in a sentence.

The outputs of these methods also differ. Semi-supervised learning produces a task-specific model, such as a classifier or regressor, tailored to the problem at hand. Self-supervised learning, however, yields pretrained representations that can generalize across multiple tasks, making it a powerful tool for transfer learning. For instance, pretrained language models like BERT or GPT, as well as computer vision models based on contrastive learning, are examples of self-supervised learning.

DOI: 10.1201/9781003561460-14

10.1 SEMI-SUPERVISED LEARNING

Semi-supervised machine learning is an approach that combines elements of both supervised and unsupervised learning to leverage both labelled and unlabelled data in the training process. Since labelling data can be expensive and/or time-consuming, semi-supervised learning is helpful in situations where obtaining large amounts of labelled data is impractical, but unlabelled data is readily available. The labelled data provides direct supervision to the model, while the unlabelled data helps it generalize better by learning from the underlying structure in the larger, unlabelled data set.

Typically, an unsupervised learning algorithm begins by training a model on a small set of labelled data using *supervised learning*. It subsequently deploys this *initial model* to make predictions on the more extensive set of unlabelled data. A set of the most confident of these predictions is assumed to be accurate and added to the labelled data set. This new, enlarged, labelled data set is subsequently used to retrain the model via supervised learning, potentially improving its performance, and the process is iterated as required.

Other semi-supervised ML methods use the structure of the unlabelled data to inform the decision boundaries of a categorical model—e.g. identifying dogs in images—helping to improve generalization. Common approaches include *self-training*, *co-training*, and *graph-based* methods. However, in all cases, the goal is to extract useful information from the large pool of unlabelled data to achieve better performance than would be possible using supervised learning on the small set of labelled data initially provided.

10.2 SELF-SUPERVISED LEARNING

Labelling every element of the world is impractical, making supervised learning, which relies on large volumes of *labelled* data, a significant bottleneck in developing machine learning models capable of diverse tasks and skill acquisition. Modern AI systems require the ability to understand their environment without being restricted to fully manually labelled data sets. This mirrors how humans and animals learn by observation, acquiring broad and generalized knowledge about the world. Recall that Yann LeCun highlighted how quickly humans learn to drive with minimal supervision due to their *understanding* of the world. In contrast, AI systems require extensive training with thousands of hours of data.

By leveraging the inherent structure and domain knowledge implicit in the data, self-supervised learning (SSL) eliminates the need for human labelling, addressing a fundamental limitation of supervised machine learning.[1] [204] Moreover, although sensu stricto, a type of unsupervised learning, SSL can bridge the gap to supervised learning by deriving pseudo-labels or structured objectives directly from the data. In this way, LeCun and others see SSL as the most promising way for AI systems to encapsulate 'background knowledge of how the world works'—*a prerequisite for recognizing and understanding patterns of subtle, less familiar features of the world*—because SSL offers a way to learn from many orders of magnitude more unlabelled data than is available from labelled sources. This approach is particularly relevant for autonomous driving, where it has recently been proposed as a predictive architecture using joint-embeddings.[2][391] We will return to this later.

However, SSL's notable downside is that while it eliminates the need for costly manual labelling, it also reduces *human oversight* of the data. This lack of oversight can have negative implications for safety-critical tasks like autonomous driving. Consequently, SSL, as deployed by AV companies like Tesla, has faced criticism, particularly for its lack of transparency and explainability—albeit a lot of this criticism has come from Mobileye, a competitor and former business partner of Tesla.

Self-supervised learning initially gained prominence in natural language processing, where it serves as the foundation for large language models such as BERT and GPT. A common approach, *masking*, involves concealing parts of the data and using the visible elements to predict the hidden ones. In NLP, this is exemplified by BERT's masked language modelling, where words or tokens in a sentence are masked.

In computer vision, masking is applied to image regions, such as patches, as seen in masked autoencoders. Similarly, for tabular data, specific columns or entries may be hidden. Across these domains, the model learns to leverage the unmasked data as context to infer the missing information, capturing relationships and dependencies within the data. Furthermore, some tasks lack suitable labelled data sets altogether, making SSL an essential approach for applying machine learning in these applications.

10.2.1 PRETEXT, AND AUXILIARY PRETEXT TASKS

Self-supervised learning employs pretext tasks to train models on unlabelled data, enabling them to learn rich, transferable representations applicable to downstream tasks such as image classification, object detection, and segmentation. These methods achieve performance comparable to fully supervised approaches.

Contrastive learning is a key technique that trains models to distinguish between similar and dissimilar image representations. Methods like SimCLR and MoCo treat augmented views of the same image—for example, rotated or cropped versions—as a positive pair to be *attracted* in the embedding space, while treating augmented views of different images as negative pairs to be *repelled.* Clustering-based algorithms, such as DeepCluster and SwAV, learn visual representations by training a model to predict pseudo-labels (*cluster IDs*). The model's representations are used to group images into clusters, and the resulting cluster assignments then serve as the training targets. This process forces the model to learn features that capture the semantic similarities between images.

Inspired by natural language processing, masked vision-prediction tasks hide portions of an image and train the model to predict the missing content. Masked autoencoders exemplify this by masking random image patches and reconstructing them. Context-based tasks, such as the jigsaw puzzle task, leverage spatial relationships by training a model to reconstruct shuffled image tiles.

Generative tasks using variational autoencoders, or GANs, require the model to reconstruct or generate parts of the input data, thereby learning underlying data distributions. In video applications, motion-based pretext tasks utilize temporal dynamics, training the model to predict the order of frames or future frames based on past frames, thereby capturing motion patterns and temporal consistency.

Auxiliary pretext tasks are secondary objectives added to the main task to improve or stabilize learning. While they also help the model learn useful representations, their role is supportive rather than central. By providing additional gradients or constraints, auxiliary tasks can enhance the performance or robustness of the model by encouraging it to capture specific data properties, such as spatial coherence, that the primary pretext task might overlook.

Notes

1. A form of self-supervised learning and the related generative adversarial networks were introduced in 1990 by Schmidhuber. [302]

2. The idea of leveraging a broader generative world model created in an unsupervised manner to generalize to unseen tasks was pioneered by David Ha and Jürgen Schmidhuber. [125]

11 New Types of Models

In this chapter, we will review some of the most prominent deep neural network models. By 2014, when Tesla entered the autonomous driving race, powerful tools for computer vision, including *structure from motion*, were already available for deployment. This knowledge sets the stage for investigating the revolutionary deep neural network architectures for autonomous driving, which are currently being deployed by companies such as Tesla and Waymo. The implications of deploying these powerful new technologies in the real world of autonomous driving will be examined in the second volume of the *Driving Intelligence* trilogy, *The Amber Book—Errors in Motion, or How Autonomy Fails*.

11.1 THE VANISHING GRADIENT PROBLEM

The backpropagation algorithm was designed to enable the training of neural networks with more than one hidden layer between the input and output layers, so building deep multilayer perceptrons with many hidden layers between the input and output should be straightforward in principle; however, a critical problem that emerges as the network depth increases is the *vanishing gradient* problem.

11.1.1 THE RECTIFIED LINEAR UNIT

Vanishing gradient occurs when the gradient of the loss function—*which gives a measure of the error between the model's predictions and the actual target values, and which is used to update the weights during backpropagation*—becomes very small as it is propagated backwards through the layers of a deep network. This causes the weights in the earlier layers (those closer to the input) to update very slowly, effectively halting learning in those layers.

Hence, vanishing gradient is a function of the network's depth and the activation function used—typically a semi-linear differentiable function such as a sigmoid. Indeed, one of the leading causes of the vanishing gradient is the use of activation functions with *limited output range*. For example, the output range of the classical sigmoid activation function is between 0 and 1.

One proposed mitigation is the use of the rectified linear unit (ReLU) activation function, which, for positive values, does not inhibit the backflow of the gradient signal. For example, observing the derivative in Figure 11.1, we notice how, for positive values, the derivative is constant, thus enabling easier backpropagation of error signals and concomitant weight updates.

11.2 THE DEEP MULTILAYER PERCEPTRON

The use of ReLU activation enables the construction of much deeper MLPs; however, as the universal approximation theorems established that feedforward MLPs with just

DOI: 10.1201/9781003561460-15

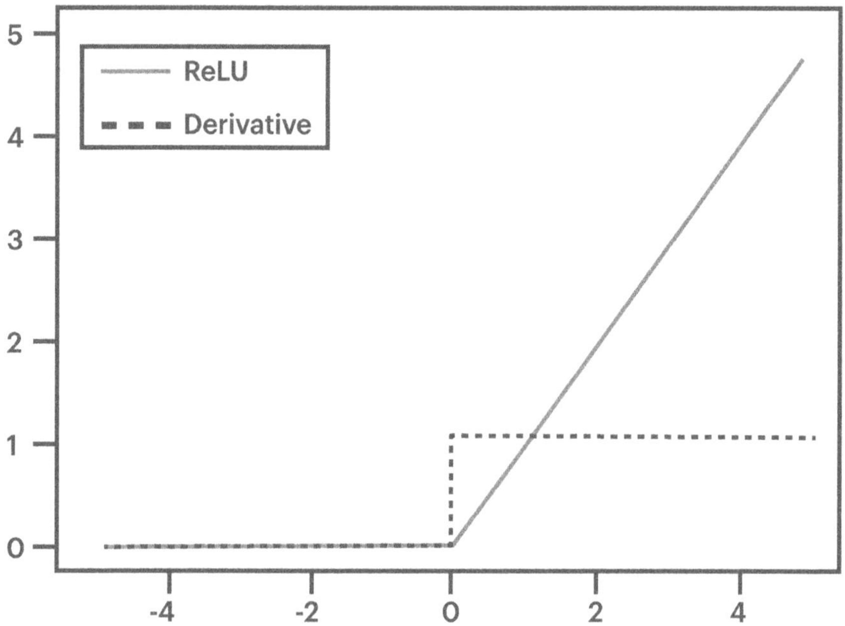

Figure 11.1 ReLU: the rectified linear unit. (Image adapted from Cannata, 2021 [55].)

a single hidden layer of finite size can approximate any continuous function (see page 81) and the addition of extra layers causes the computational cost of the network to increase significantly, what are the benefits of going deep in network design?

Well, while it is true that, in theory, an MLP with a single hidden layer can approximate any continuous function, in practice, deep neural networks can provide several critical advantages over shallow ones:

Improved representational power: DNNs can represent highly complex and abstract features that shallow networks cannot capture. With more layers, the network can learn more complicated and abstract representations of the input, which can lead to better generalization and more accurate predictions.

Robustness to noise: DNNs are less sensitive to noise in the data than shallow networks. The reason is that deeper networks can learn to filter out noise at different levels of abstraction, whereas shallow networks typically merely filter noise at the input level.

Faster training: DNNs can often be trained faster than shallow networks. This is because the lower layers of the network can learn basic features, which can be reused in different parts of the network. This reduces the overall number of parameters that need to be trained.

Transfer learning: DNNs can be used for transfer learning, which is the process of leveraging pretrained models for new tasks. This is possible because the deeper layers of the network can learn generic features that can be useful for many different tasks.

Hence, while shallow neural networks can approximate any continuous function, in practice, DNNs provide many benefits that can lead to better performance and more efficient learning.

11.3 CONVOLUTIONS AND DEEP NETWORKS

The hegemony of radial basis function, support vector machines, and Bayesian methods in machine learning conferences began to wane as new convolutional deep learning algorithms were developed and began to be applied in computer vision, as paradigmatically demonstrated by the success of AlexNet—Alex Krizhevsky, Ilya Sutskever, and Geoffrey Hinton's entry to the 2012 ImageNet competition (see Section 11.3.3). However, the evolution of convolutional approaches to image processing dates back much further, to 1980, with Kunihiko Fukushima's neocognitron.

11.3.1 THE NEOCOGNITRON

The neocognitron, developed by Kunihiko Fukushima in 1980, [107] was an early multilayer, unsupervised, artificial neural network inspired by the seminal work of David Hunter Hubel and Torsten Wiesel on the visual pathway of cats and macaque monkeys.[1] Its architecture is comprised of alternating layers of S-cells (simple cells) and C-cells (complex cells), mirroring the function of simple and complex cells in area V1 of the mammalian visual cortex. [156]

The first layer of S-cells contains carefully specified, heterogeneous filters resembling Gabor functions,[2] analogous to the simple cells in V1. These filters, sensitive to edges and contrasts at various orientations, are each *convolved* across the entire input image.

Convolution can be thought of as incrementally sliding a *filter*, analogous to a lens, over small areas of the input image to produce a *feature map* that highlights specific filtered patterns or features, such as edges, textures, or shapes. If the convolutional filter has a receptive field of, say, $x \times y$ pixels, then the filter is defined by a $x \times y$ matrix of real-valued filter coefficients (*weights*). Hence, by using k first-layer filters, each point in the convolution generates a k-dimensional output, so defining k feature maps of the input image.

The subsequent C-cell layer performs a form of local pooling on these feature maps, *downsampling* the image by *spatially averaging* the activation of S-cells lying within their receptive fields. This introduces a degree of positional invariance and reduces the spatial dimension of the vector formed by the outputs of the C-cells in a process akin to that performed by the complex cells of V1.

Learning in the neocognitron primarily occurs in the adaptation of the connection weights between S-cells and C-cells, utilizing a *Hebbian-like* unsupervised learning rule, i.e. when an input pattern is presented, the S-cell with the *most vigorous* response in a local group has its connections to the corresponding C-cell *strengthened*, reinforcing the network's ability to detect specific features. This unsupervised *competitive learning* mechanism allows the neocognitron to adapt its feature detectors without the use of modern ML methods such as backpropagation.[3]

As processing continues through subsequent layers, the output of each layer is reduced in size by C-cell downsampling, and the complexity and variety of the S-cell filters increase, as does the size of their [effective] receptive fields. In this way, the higher layers of the neocognitron combine features from previous layers to detect ever more intricate patterns, evolving from simple edge detection to complex shape recognition.

The final layer of the neocognitron, consisting of a set of S-cells with receptive fields spanning the entire previous output vector, represents the network's interpretation of the input image. The S-cell that responds most strongly to the input image represents the class identifier for the neocognitron's input image.

This hierarchical structure enables the neocognitron to build increasingly abstract representations of the input image, laying the groundwork for modern convolutional neural networks.

In summary, we can think of the neocognitron functioning a bit like a detective agency, where different teams analyse clues at various levels of complexity. The first team looks for basic elements (edges, corners), the second team combines those clues into more recognizable patterns (shapes), and the final team uses all the compiled information to solve the entire mystery (recognizing objects or scenes). This mirrors how the neocognitron extracts increasingly complex features from an image, moving from simple to complex recognition.

Subsequently, in a variant of the neocognitron called the *cresceptron*, introduced by Weng et al. in 1992, instead of using Fukushima's *spatial averaging* for downsampling, a new method called *max-pooling* was introduced. This development remains widely used in modern convolutional approaches.[4] During max-pooling, the input map is divided into a set of non-overlapping rectangular regions called 'pooling windows'. For each window, the maximum value in that region is computed and placed into the output map; the other values in the window are discarded.

11.3.2 THE CONVOLUTIONAL NEURAL NETWORK

The first impact of modern convolutional neural network (CNN) approaches in computer vision tasks is usually positioned with Yann LeCun's *LeNet-5* program, as described in his 1998 paper 'Gradient-Based Learning Applied to Document Recognition'. [202]

Specifically designed to recognize handwritten digits, LeNet-5 demonstrated the potential of CNNs on practical vision tasks, where it achieved impressive results on the MNIST handwritten digit data set.[5] It marked a significant milestone in the subsequent development of CNNs, paving the way for further research and applications in computer vision. However, the roots of CNNs go back to 1987 and the work of Homma et al. on phoneme recognition, [148] which introduced the term *convolution* in this context.[6]

Compared to the earlier neocognitron, LeNet-5 proved to be a significant architectural development. It introduced vital innovations, including a deeper structure, backpropagation training, a sigmoid activation function and adaptive convolutional filters.

Unlike the carefully manually specified convolutional filters of the neocognitron, LeNet-5 deployed filters with *learnable* weights, which allowed the network to process the most relevant features from the input data *automatically* (see Figure 11.2). However, like the neocognitron, LeNet-5 deployed *spatial pooling*, aka *subsampling*, to downsample feature maps, reducing computational costs and enhancing the network's invariance to small translations and rotations. Finally, fully connected MLP layers were used to combine the extracted features for classification.

Although LeNet-5 successfully introduced convolutional and spatial-pooling layers for feature extraction and dimensionality reduction, it was limited in depth and computational efficiency. However, in 2012, as we observed in Section 9.3, AlexNet was to address these limitations and demonstrate the true power of deep learning for image recognition tasks.

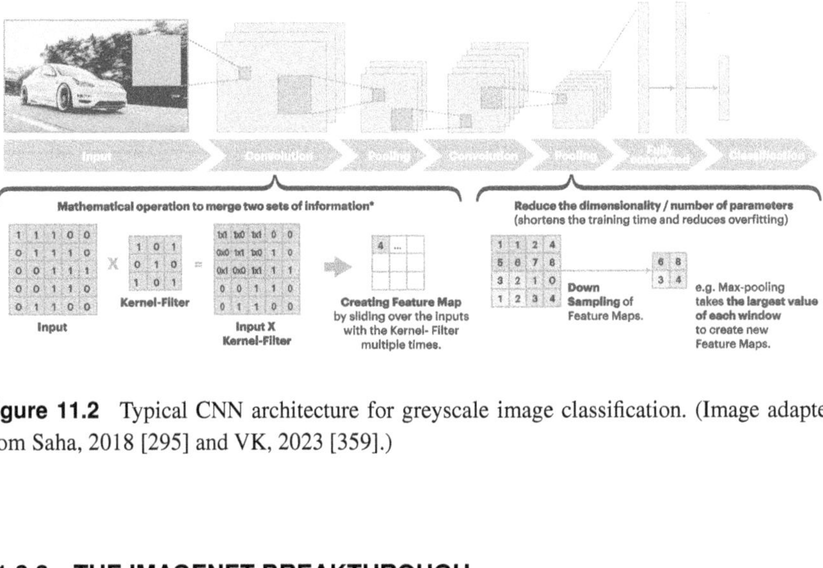

Figure 11.2 Typical CNN architecture for greyscale image classification. (Image adapted from Saha, 2018 [295] and VK, 2023 [359].)

11.3.3 THE IMAGENET BREAKTHROUGH

The ImageNet breakthrough occurred when a newcomer team introduced a novel form of *deep convolutional network*, achieving a significant step change in image recognition and classification performance that, in retrospect, can be seen as a revolution.

AlexNet, a program named after Alex Krizhevsky, the first author in the paper that made it famous, achieved a top-5 error rate of 15.3%, meaning that the correct class was not in the top 5 predicted classes 15.3% of the time. [190] This performance marked a significant improvement over the second-best entry, which had a top-5 error rate of 26.2%. Furthermore, it marked an improvement of 1,040 basis points over the previous year's winner, which had achieved a top-5 error rate of 25.7%.[7]

AlexNet's success was essentially based on two pillars: it had been trained on a vast amount of data, and its architecture was such that only a few parameters needed to be tuned. At the heart of AlexNet are its convolutional layers, which extract features from the input image by applying *learnable* filters—i.e. at each position in the convolution, the filter applies an *activation function* to the weighted sum of pixel values from the image, multiplied by the corresponding filter weights.

In AlexNet, similar to LeNet-5 and in contrast to the neocognitron, the weights defining the filters are *learnt* during training, allowing the network to automatically discover the most informative features for the task at hand. But, unlike LeNet-5, to introduce non-linearity and avoid the vanishing gradient problem, AlexNet employs the rectified linear unit (ReLU) non-linear activation function, as described in Section 11.1.1, setting negative activations to zero while outputting positive values unchanged.

Subsequently, local response normalization (LRN) was deployed as a technique to improve the generalization performance by encouraging *lateral inhibition*, analogous to the biological process where excited neurons suppress the activity of neighbouring neurons. LRN was particularly effective in combination with ReLU activation functions, as it mitigated the high variability in activation magnitudes caused by ReLU by downscaling large activations within its local neighbourhood.

Furthermore, in AlexNet, the number of filters increases in deeper layers, enabling the network to learn more complex and abstract features. Thus, AlexNet typically deploys 96 filters in the first layer, followed by 256, 192, 256, and 256 in the subsequent layers.

Similar to Weng's 1992 cresceptron, AlexNet also deploys max-pooling to preserve the essential information while downsampling the dimensionality of feature maps. A process which can further help prevent overfitting and minimize computation costs. Additionally, after max-pooling, the network became tolerant of small translations in the original input image.

The final layers of AlexNet are fully connected, similar to those in an MLP, and combine the extracted features into a predicted class. Indeed, both the weights and filter parameters in AlexNet are trained using supervised learning via backpropagation. Hence, in order to learn the relationship between input images and their corresponding classes, the training set must consist of fully *labelled* data.

To further prevent overfitting, AlexNet additionally used *dropout regularization*. Here, during each iteration of backpropagation learning, a random subset of neurons in a layer is temporarily removed, i.e. dropped out or set to zero, such that they do not contribute to the forward or backward pass of the algorithm. Such regularization effectively forces the network to not rely too much on any one neuron, promoting redundancy and better *generalization* to unseen data.

Thus, several innovations contributed to AlexNet's success. The architecture was deep, consisting of 8 layers of learned weights (five convolutional layers and three fully connected MLP layers). ReLU activations were employed to address the vanishing gradient problem, enabling faster training of deeper networks. Local response normalization was introduced to encourage competition between neurons and improve generalization. Additionally, overlapping max-pooling was used to reduce spatial dimensions while retaining essential information. AlexNet also utilized data augmentation techniques, such as random cropping and mirroring, to prevent overfitting, along with dropout regularization.

Following the success of AlexNet, deep learning rapidly became the dominant approach in subsequent ImageNet challenges, both in academia and industry, where it continued to reduce top-5 error rates significantly. For example, in 2015, Microsoft Research's *ResNet*—short for *Residual Networks*—achieved a top-5 error rate of 3.57%, which brought the error rate very close to human-level performance.[8]

Autonomous driving would not be possible without translational invariance, which means that the model can recognize objects regardless of their position in the image, for example, allowing it to detect the same features whether they appear in the centre, corner, or anywhere else. Although CNNs like AlexNet are not fully invariant to translations, they have enabled autonomous vehicles to operate in more complex scenarios that require accurate object detection and classification. We will return to this success later in our story.

11.4 FIRST STEPS IN GENERATIVE AI

11.4.1 THE BOLTZMANN MACHINE

The Boltzmann machine, invented by Geoffrey Hinton and Terry Sejnowski in 1985, is a type of stochastic neural network designed to model probability distributions over a set of binary variables. [140] Named after the physicist Ludwig Boltzmann, the machine uses concepts from statistical mechanics, specifically the Boltzmann distribution, to adjust its weights and learn patterns from data.[9]

Boltzmann machines are best suited for problems like combinatorial optimization, where the goal is to find the best solution among many possible ones, such as travelling salesman problems or optimising resource allocation. They are also used in unsupervised learning to discover complex structures in data, such as dimensionality reduction or feature learning.

The Boltzmann machine consists of a network of simple units (neurons) that can be in one of two states: On (1) or off (0). These units are connected, and each connection has a weight that influences the interaction between units. The system learns by adjusting these weights to minimize its overall energy.

The learning process involves matching the patterns generated by the machine to those found in the training data. This is done in two phases. In the *positive phase*, the machine is presented with data and adjusts its weights to reduce the energy for patterns resembling the input data. In the *negative phase*, the machine runs without input and adjusts the weights to increase the energy of patterns that it generates, but which do not match the data. By iterating these phases, the Boltzmann machine learns to create patterns similar to those in the training data.

In operation, classical Boltzmann machine learning is akin to *simulated annealing*. [184] In annealing, a material is heated to a high temperature and then slowly cooled. At high temperatures, the atoms of the material have enough energy to explore many configurations, even those with high energy. As the temperature decreases, the particles settle into configurations with lower energy, eventually reaching a stable, low-energy state. The slow cooling process ensures that the system avoids becoming trapped in local minima and instead converges towards a global minimum. Unfolding the analogy, we observe:

- Both the Boltzmann machine and simulated annealing are grounded in concepts from statistical mechanics, particularly the Boltzmann distribution. In both techniques, a *temperature* parameter controls the stochasticity of the system. In simulated annealing, the temperature is gradually lowered to find optimal solutions, while in Boltzmann machines, it can be used to guide the learning process.
- Both methods aim to minimize an energy function. In Boltzmann machines, this energy function relates to the configuration of the network, whereas in simulated annealing it represents the cost or objective function of the optimization problem. Probabilistic transitions between states are central to both techniques: In Boltzmann machines, neurons stochastically decide to be on or off, and in simulated annealing, transitions to new states are accepted probabilistically based on the energy difference and the temperature.

Boltzmann machines can be considered an early form of *generative AI* due to their ability to model complex probability distributions over binary random variables and generate new data samples based on statistically learned patterns.[10] The central idea is that Boltzmann machines *learn the underlying distribution of a data set* and can then use this knowledge to generate new data that resembles the original data. However, classical Boltzmann machines are computationally expensive and challenging to train for large-scale problems, so they are not widely used today in comparison to other ML models.

11.4.2 THE RESTRICTED BOLTZMANN MACHINE

A more practical form of the Boltzmann machine, used for generative tasks, is the restricted Boltzmann machine (RBM). [329] RBMs simplify the complete Boltzmann machine by limiting the connections to only between visible units (representing data) and hidden units (representing latent features), with no intra-layer connections. Furthermore, unlike classical Boltzmann machine learning, RBMs employ different techniques for training and sampling that are more efficient and easier to train—methods such as *contrastive divergence* and *Gibbs sampling*, rather than *simulated annealing*.

Indeed, a form of RBM has been successfully deployed in the context of autonomous driving, where it was proposed as a method to detect road accidents reliably, [283]—a problem which can cause severe issues for AVs, as we highlight in our second volume, *The Amber Book: Errors in Motion, or How Autonomy Fails*.

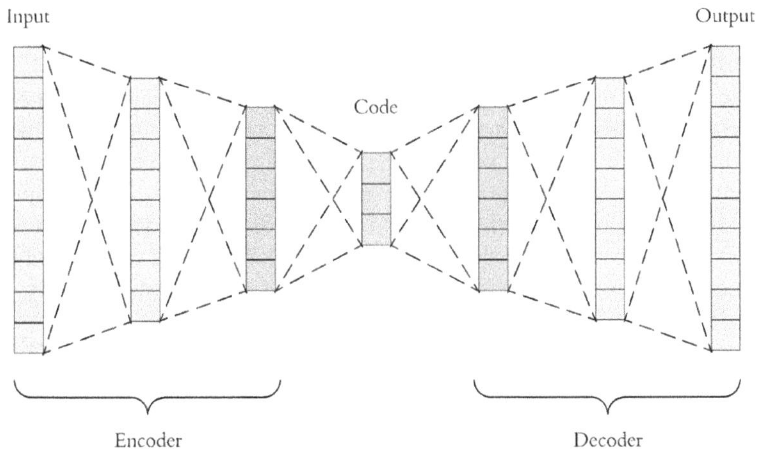

Figure 11.3 The deep autoencoder network. (Image adapted from Dertat, 2017 [84].)

11.5 AUTOENCODERS, DEEP BELIEFS AND GANS

In 1986, Ralph Linsker, an IBM scientist, published information on an early form of unsupervised learning in neural networks. [212] Linsker's work explained how a neural network could self-organize and develop structured features without explicit supervision, drawing inspiration from biological systems, particularly the visual cortex. The central idea was to show that simple Hebbian learning rules, combined with a random initial configuration could lead to the emergence of organized structures, such as edge detectors, similar to those found in biological visual systems.

At the heart of his research, Linsker's *infomax principle* suggests that neural networks organize themselves to maximize the mutual information between inputs and outputs, so learning *compact* representations of data, which preserve essential information.

As we will subsequently see, Linkser's concept of constraining information flow through the network to force it to learn efficient representations would become the central guiding principle in *autoencoder* design.

11.5.1 AUTOENCODER NETWORKS

Deep autoencoders (AEs) are a type of unsupervised feedforward multilayer perception where the input is identical to the output. They function by compressing the input into a lower-dimensional representation, known as the *latent space* or *code*, and then reconstructing the original input from this compressed form. Hence, the code functions as a *compact representation* of the input. In operation, an autoencoder network is given a data input x, which it maps to a code, a latent representation, z, from which the decoder network reconstructs the data input x.

Thus, a simple AE consists of three components: an *encoder*, *code*, and a *decoder*. The encoder compresses the input to generate the code, while the decoder reconstructs the input defined by the code.

Simple AEs are often used for dimensionality reduction, albeit with particular characteristics, i.e. they are *data-specific*, meaning they can only compress data similar to what they were trained on. Unlike traditional compression algorithms like GZIP, autoencoders learn features *specific* to their training data. For example, an autoencoder trained on handwritten digits will not effectively compress landscape photos. Furthermore, they are *lossy*, meaning the output will be a close but not perfect replica of the input, making them unsuitable for applications requiring ideal compression.

A simple form of autoencoder uses fully connected feedforward MLPs for both the encoder and decoder. The code is implemented using a single-layer feedforward neural network with n neurons, and this hyperparameter—defining the code size—is set before training. Typically, the decoder mirrors the encoder in structure, though this is not a strict requirement; the only necessity is that the dimensionality of the input and output vectors match.

Four hyperparameters must be set before training an autoencoder: the code size, which controls the degree of compression; the number of layers in both the encoder and decoder; the number of nodes per layer; and the loss function. For inputs in the range [0, 1], a binary cross-entropy loss function is typically used.[11] Alternatively, a simple mean square error loss function can be applied.

In this simple AE, as described above, training is typically performed via backpropagation, as is standard for MLPs. For valid inputs, this compresses the data by eliminating redundancy through the encoder. However, because the latent space in the AE is not regularized (see the *white* areas in Figure 11.4), the decoder cannot generate valid input data from all vectors randomly sampled from the latent space; hence, this simple AE is not, sensu stricto, a fully *generative* system.

11.5.2 THE VARIATIONAL AUTO-ENCODER

The variational autoencoder (VAE) addresses the issue of the non-regularized latent space in standard autoencoders by introducing *generative* capability across the entire latent space. Instead of outputting latent vectors directly, the VAE's encoder generates *parameters* for a pre-defined probability distribution in the latent space. This distribution is constrained to be *Gaussian*, thereby regularizing the latent space and ensuring smoother sampling.

During training, the input data x is passed through the encoder, represented as $e_\theta(x)$, which reduces its dimensions to create a compressed latent representation. However, rather than outputting the latent vector directly, the encoder provides the mean and standard deviation for each latent variable. The latent vector z is then sampled from this distribution and fed into the decoder, which reconstructs the input in a manner similar to a standard autoencoder.

The VAE's loss function incorporates two objectives: reconstructing the input and ensuring that the latent space follows a Gaussian distribution. The total training loss

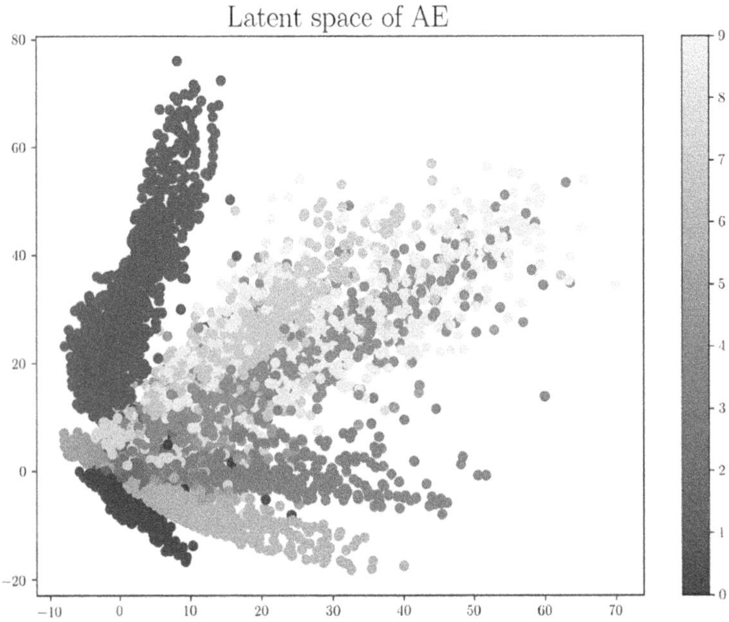

Figure 11.4 The latent space of a simple autoencoder. (Image adapted from Anwar, 2021 [13].)

is the sum of the reconstruction loss, which measures the difference between the input and reconstructed output, e.g. using mean squared error, and the similarity loss, which is the Kullback–Leibler divergence between the latent space distribution and a Gaussian distribution.[12]

Since random sampling from the encoder's distribution complicates backpropagation, the VAE uses a *reparameterization* trick. This technique expresses the latent vector z as a function of the encoder's output, allowing gradients to propagate through the network and enabling practical training.

In the VAE, because the latent space of the VAE *is* regularized (see Figure 11.5), the decoder *can* generate valid input data from random samples of the latent space; hence, the VAE can function as a fully *generative* system. For example, in autonomous driving, VARs have been successfully applied to vehicle path prediction. [163]

11.5.3 DEEP BELIEF NETWORKS

Densely connected networks, also known as deep belief networks (DBNs), are typically used for unsupervised learning tasks, such as feature extraction, dimensionality reduction, and generative modelling. In 2006, Geoffrey Hinton and colleagues described DBNs based on layers of restricted Boltzmann machines and contrastive

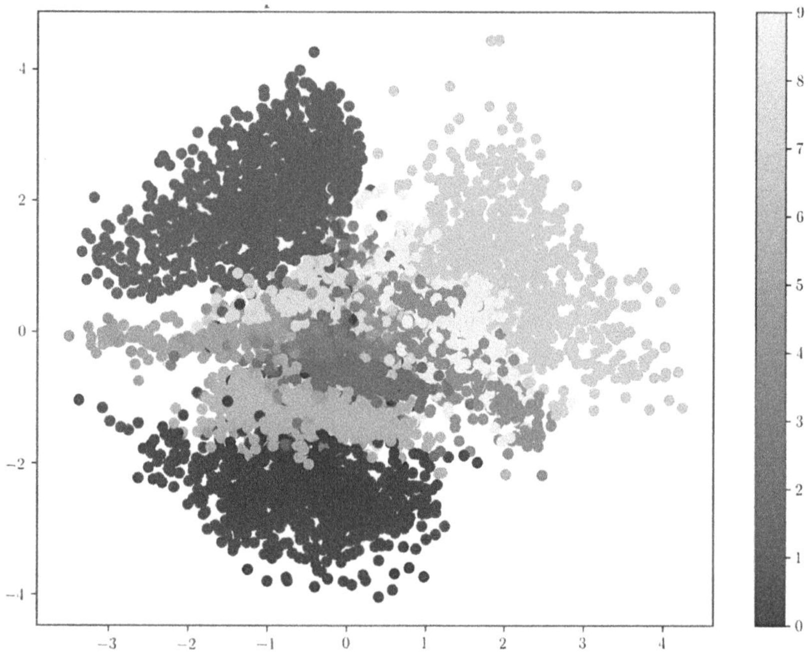

Figure 11.5 The latent space of a variational autoencoder with KL loss. (Image adapted from Anwar, 2021 [13].)

divergence learning. [139] This was followed, in 2007, by Bengio and LeCun, who published variants based on stacks of autoencoder networks.[13]

A DBN is a class of generative DNNs composed of multiple layers of latent variables (hidden units), with connections between the layers but not between units within each layer. When trained on a set of examples without supervision, a DBN can learn to probabilistically reconstruct its inputs, with the layers acting as a stack of hierarchical feature detectors. However, DBNs can also be used for supervised learning tasks, such as classification.

For use in classification, after learning the generative model of the training data, the DBM can be further trained through supervised learning to map the learned representations to the output classes, thereby performing *generative discrimination* (classification). This enables the DBN to learn more complex and abstract representations of the input data, thereby improving the accuracy of the classification.

Consider the pattern recognition performance of a DBN network with three hidden layers and about 1.7 million weights on the MNIST set of handwritten digits, as reported by Hinton et al. (2006). [139] After training on 60,000 images with no knowledge of image geometry or special preprocessing, the generalization performance of the network is 1.25% errors on the 10,000-digit official test set. This surpasses the 1.5% achieved by the best backpropagation nets when they are not explicitly designed for this particular application.

Overall, one of the key advantages of DBNs is that they can utilize unsupervised learning to derive *hierarchical representations* of the input data, where each layer learns to represent increasingly abstract and high-level features. For example, DBNs can be considered a stack of simple, unsupervised *autoencoders*, where each sub-network hidden layer serves as the visible layer for the next and can be trained, AE layer by AE layer, to develop a hierarchical representation of the latent variables representing the training data. This can be particularly useful for tasks such as image recognition, speech recognition, and natural language processing, where the input data has a hierarchical structure.

11.5.4 GENERATIVE ADVERSARIAL NETWORKS

In 2014, Ian Goodfellow et al. described an exciting new type of ML architecture called 'generative adversarial networks' (GANs). [118] GANs deploy two *adversarial* neural networks: One, the *generator*, synthesises new data instances, while the other, the *discriminator*, rates each instance as how likely it is to belong to the training data set.[14]

Colloquially, the *generator* takes the role of a counterfeiter and the *discriminator* the role of the police in a complex and evolving game of cat and mouse, wherein the counterfeiter is evolving to produce better and better counterfeit money while the police are getting better and better at detecting it. This game continues until, at convergence, both networks have become highly proficient at their tasks.

In their paper, Goodfellow et al. describe using a feedforward multi-layer perceptron network with a combination of rectified linear and sigmoidal units for the activation functions of the generator network. In contrast, the discriminator network uses a combination of *maxout* activations and dropout regularization. [119].

In a *maxout layer*, instead of applying a standard activation function (like ReLU or sigmoid), the output is the maximum value from a set of linear functions. Each neuron in a maxout layer computes the maximum of several affine transformations—i.e. linear functions—of the input. Their approach yields good results on both the MNIST and Toronto face data sets.

Several attempts have been made to combine GANs with autoencoders. Larsen and Sønderby introduced a variational autoencoder that alternated its training with a generative adversarial network setup, where the decoder of the VAE served as the generator in the GAN architecture. [195] This approach enabled the generation of face samples with more detailed high-frequency content than a standard VAE, which tended to produce smooth outputs.

Makhzani et al. adopted a similar strategy, using the GAN to shape the latent space based on classification labels. Their method allowed the latent space to be directly transformed into Gaussian distributions or more complex structures, such as a *Swiss roll* distribution. [220]

One notable example which illustrates the generative reconstructive power of VAEs combined with GANs is from Terrence Broad. [47] On the right-hand side of Figure 11.6, we see the output of such a system designed to process Blade Runner—*a well-known sci-fi film which riffs on the notion of what is human and what is*

Figure 11.6 Terrence Broad's autoencoding network 'dreams' of Blade Runner. (Image adapted from Broad, 2016 [47].)

machine—building up its internal representations of that film and then re-rendering these to produce an output movie that is surprisingly similar to the original (shown on the left).

In Broad's dissertation, he describes encoding/reducing each frame of Ridley Scott's Blade Runner to a 200-digit vector before invoking a *decoder* to reconstruct each frame just using those numbers. The result appears akin to an android's dream—the neural net, working without human supervision to identify significant features in each frame, was able to capture the essential elements so well that its reconstruction of Blade Runner triggered a *copyright takedown notice* from Warner Brothers when it was uploaded to YouTube.

11.5.5 APPLICATIONS OF GENERATIVE ARTIFICIAL INTELLIGENCE

Generative AI has been used to perform many important functions, particularly in data generation and representation learning. One of its primary uses has been *data generation*, as it can create new samples that resemble the training data by sampling from its regularized latent space. This capability is valuable for generating images, text, or other forms of structured data.

Generative AI has also proven effective for *dimensionality reduction*, compressing high-dimensional data into a lower-dimensional latent space while preserving important features. Such compression is particularly useful in tasks like *image reconstruction*, where the model can reconstruct images from their compressed latent representations. This ability extends to applications such as *image denoising* and *inpainting*, where parts of an image are restored or completed.

Similarly, variational autoencoders can be applied to *data imputation*, filling in missing parts of data sets, such as completing incomplete images or time-series data. In addition, VAEs have been successfully applied to problems, including *anomaly detection*, where they can identify outliers by flagging data points that do not fit the latent Gaussian distribution, making VAEs useful in applications such as fraud detection. Moreover, the regularized and continuous latent space in VAEs allows for *latent space exploration*, enabling smooth transitions between data points, such as interpolating between images. Furthermore, VAEs can be powerful tools for *representation learning*, capturing meaningful latent features that can be leveraged for tasks such as classification or clustering, where understanding the underlying patterns in the data is essential.

Generative adversarial networks have been successfully used for a variety of tasks, such as image synthesis, style transfer, and data augmentation. One of the most well-known applications of GANs is in computer vision, where they have been used to generate realistic images of objects, landscapes, and even people. GANs have also been deployed in natural language processing tasks such as text generation, language translation and summarization.

GANs have also been utilized in scientific research, such as generating new molecules for drug discovery or creating realistic simulations of physical systems for scientific experimentation. In addition, GANs have been used in the field of art to develop new and exciting digital artworks (cf. Terrence Broad's version of an *android dreaming*. [47]).

In autonomous driving, generative AI is currently being explored in various applications, including simulating driving scenarios for decision-making, map generation, simulation and testing, and developing new natural language interfaces. Indeed, at NIPS 2016, not long before the emergence of large language models and ChatGPT, Yann LeCun, Facebook's AI Director of Research, famously claimed *generative adversarial networks* to be 'the most interesting idea in machine learning in the last ten years.'[15]

Notes

1. The sequence of neural processes and structures in the brain responsible for visual perception, specifically the visual information processing apparatus linking the retina to the primary visual cortex, V1.

2. Gabor functions, or Gabor filters, are commonly used in convolutional filters for tasks involving image processing and feature extraction. They are named after Dennis Gabor, who originally introduced them to model human visual perception. Gabor filters are particularly well-suited for capturing specific types of patterns, such as edges, textures, and frequency information in images, which makes them useful in computer vision. [108]

3. Competitive learning is a type of unsupervised learning in neural networks where neurons (or units) in the network compete to become active in response to a given input. In this framework, only a small subset of neurons 'win' the competition and update their weights, while the other neurons remain unchanged. The idea is to encourage the network to specialize different neurons to respond to specific patterns or features in the data, allowing the network to discover structure within the input data without explicit labels.

4. The term 'max pooling' was first formally introduced in 1990 by Yamaguchi et al. They combined time

delay neural networks (TDNNs) with max pooling to create a speaker-independent isolated word recognition system. [388]

5. A modern implementation of LeNet-5 using Keras on the MNIST data set achieved an accuracy of over 99% on the test set. [268]

6. The first paper identifying convolutional networks was published in 1989 by Alex Waibel, who achieved shift-invariance for phoneme recognition, [364] alongside two papers from LeCun et al.: *Generalization and Network Design Strategies* [197] and (ii) *Backpropagation Applied to Handwritten Digit Recognition.* [201] NB. As we have seen, the core of the CNN architecture was first introduced by Kunihiko Fukushima with the neocognitron at a time when no adequate computing power was available to implement it. [107] Alex Waibel combined ANNs with *convolution*, *weight sharing* and Seppo Linnainmaa's *reverse mode of automatic differentiation* from 1970. [364] [213] In addition, Zhang et al. had the first backpropagation-trained two-dimensional CNNs, but it took Yann LeCun to integrate these ideas and implement them on modern compute.

7. AlexNet also excelled in top-1 accuracy, although the top-5 metric is more typically used as the key performance indicator in this competition.

8. ResNet can be viewed as an open-gated variant of the earlier *Highway Net* by Rupesh Kumar Srivastava, Klaus Greff and Jürgen Schmidhuber, as setting the gates of a Highway Net to [1.0] effectively implements a ResNet. [332]

9. A similar model, based on simulated annealing, was described some years earlier by David Sherrington and Scott Kirkpatrick in 1975. [321]

10. Arguably, an even earlier form of generative AI were hidden Markov models.

11. Cross-entropy loss is a measure used to quantify the difference between two probability distributions. In machine learning, it is commonly used to evaluate how well a model's predicted probabilities match the actual labels. For classification tasks, cross-entropy loss increases as the expected probability diverges from the actual label. A lower cross-entropy loss indicates better performance, meaning the model's predictions are closer to the actual labels.

12. Kullback–Leibler divergence is a way of measuring how different one probability distribution is from another. It gives a sense of how much information is lost when one distribution is used to approximate another.

13. Cf. Bengio et al., (2007), [25] and LeCun and Bengio (2007). [199] NB. Jürgen Schmidhuber suggests that the essential foundations date back to the 1990s. [304]

14. Jürgen Schmidhuber suggests that these ideas can be traced back to the notion of *artificial curiosity and boredom* or *prediction minimization.* [307]

15. Albeit Jürgen Schmidhuber argues that sensu stricto, he originally came up with the core idea—*adversarial and generative networks*—twenty-five years earlier. [302] [305]

12 A Whole New Industry Unfolding

The 2007 Urban Challenge marked the final DARPA event focused on self-driving cars, as the 'US Department of Defense' shifted its research priorities.[1] With DARPA no longer at the helm, the engineers involved in these challenges had to seek alternative funding sources to continue advancing their technology.

The second DARPA Grand Challenge sparked the broader application of machine learning in AV, while the third concentrated on integrating and commercializing successful methods. This trend persisted in the new non-university environment: The approaches that excelled in the Urban Challenge were further refined, but the emphasis shifted to deployment, operation, and business models. Autonomous driving was largely considered a solved problem, and the focus turned to scaling it. By 2018, over 263 companies were working on autonomous vehicle technology, according to Wired magazine, with ABC News reporting that 52 companies had been approved to test self-driving cars in California alone. [282]

The potential for a breakthrough in autonomous driving, which could fundamentally disrupt the world's second-largest industry segment, has attracted unprecedented funding. Already by 2019, traceable investments in the sector had reached approximately $166 billion, based on publicly disclosed transactions. This rapid investment surge demonstrated that the DARPA challenge ecosystem had become the foundation for a burgeoning new industry.

Many of the key players in DARPA competitions went on to lead ventures within significant companies vying for dominance in the nascent self-driving market. This remarkable industry evolution has been thoroughly documented in books by former GM R&D head Larry Burns and Wired journalist Alex Davies. Through extensive interviews and oral histories, these works provide a detailed account of how DARPA Challenges paved the way for today's autonomous driving technologies.

With the advent of machine learning, the race to revolutionize the automotive industry was fully underway. To better understand the world of autonomous vehicles, it's essential to examine what defines them and the levels of capability they must achieve to become truly autonomous.

12.1 DEFINING AUTONOMOUS DRIVING

Elaborate definitions of autonomous driving were not necessary during the DARPA Challenges. Despite their different capabilities, all the vehicles shared one commonality: they did not require a driver to operate them. As automation approaches have grown more diverse, we have to take a more differentiated approach.

Even among experts, confusion persists regarding the various levels of autonomy, despite the establishment of internationally accepted standards. The most commonly

DOI: 10.1201/9781003561460-16

used definition was developed by the American Society of Automotive Engineers (SAE). The respective norm, SAE J3016, was first published in 2014 and updated in 2016, 2018, and 2021 (the latter in collaboration with the International Organization for Standardization—ISO). The United Nations (UN) adopted this definition, which is now widely used globally.

However, the SAE was not the first to tackle the issue of categorizing autonomous driving. The original version of the taxonomy was published in 2012 by the Federal German Highway Research Institute, BASt (Bundesanstalt für Straßenwesen). [111] BASt proposed five levels, ranging from 'driver only' to 'full automation'. Notably, even at the most advanced level, it was still assumed that a driver would be present behind the wheel.

A year later, the National Highway Traffic Safety Administration (NHTSA) in the United States introduced a similar classification system. [258] [259] Building on the BASt and NHTSA frameworks, SAE introduced its taxonomy with several significant modifications. [333] SAE expanded the classification to six levels (including a Level–0), with the addition of Level–5 being the most significant change. While Level–4 in the SAE system refers to restricted automation, Level–5 signifies full autonomy without *any* limitations. This distinction, dividing NHTSA's Level–4 into two separate levels was a critical development, but it was also the most misunderstood aspect.

The confusion arises because Level–5 represents not just a technological progression from Level–4 but a conceptual one. It envisions vehicles capable of autonomous operation in *all* conditions where a human driver could operate. This is often overlooked in market studies that project Level–5 cars within a few years without addressing whether such vehicles can ever be built.

Even today, ten years after SAE's definition was introduced, experts often mistakenly equate Level–5 with an enhanced Level–4 or assume that driverless robotaxis qualify as Level–5. In reality, the critical feature of Level–5 is not the absence of a driver, but the vehicle's ability to perform autonomously in every situation a human driver could face—a goal that hinges on achieving human-level AI, i.e. engineering a form of AGI.

And yet the term 'all conditions' itself remains vague, as even human drivers cannot drive in *all* conditions.[2] Thus, SAE clarifies that *all conditions* refers to *all driver-manageable road conditions*—any time, any place, a human driver could drive (see Figure 12.1).

The presence or absence of pedals and a steering wheel did not justify a distinct level in SAE's classification. In fact, SAE did not consider it necessary to provide a specific explanation for this. It is only briefly mentioned in the description of a Level–4 vehicle that these controls may or may not be included. The shift from a driveable car to a fully autonomous pod occurs somewhere within Level–4, but for the definition, it is irrelevant whether a human can drive the vehicle or not. In the sense of SAE, the autonomous robots of all DARPA Challenges—however different in capability—were Level–4 cars in the same way as sophisticated inner-city commercial robotaxis from Waymo.

But let's take one step back and view this statement in a broader context: Automation levels are defined by a critical distinction—whether the *driver* or the *system* is responsible for the driving task. Until 2021, international law placed this responsibility solely on the driver. Self-driving vehicles were permitted only in limited test areas, subject to special exemptions. It was not until January 2021 that the UN began implementing regulations allowing drivers to temporarily transfer driving responsibility to an automated system under specific conditions. This marked a departure from a fundamental rule of the UN's Vienna Convention on Road Traffic, which requires a driver to be responsible for the entire driving process—what SAE refers to as the *dynamic driving task* (DDT). [353]

SAE engineers emphasised that the driving task, not the vehicle itself, is automated. The DDT is precisely defined to include fallback performance and minimal risk conditions.

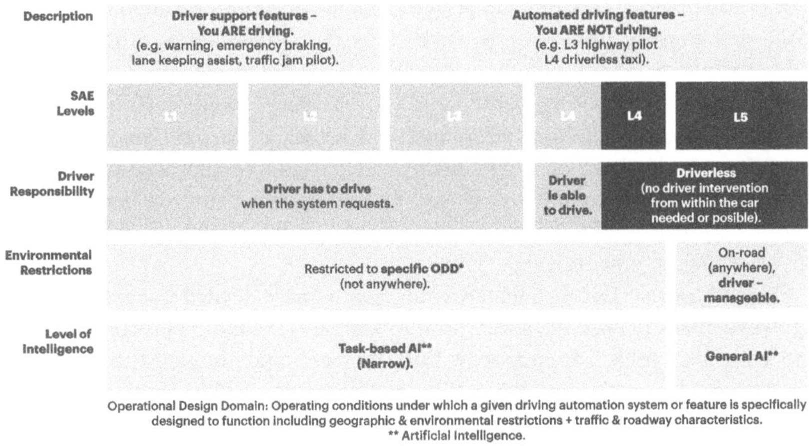

Figure 12.1 Levels of vehicle autonomy as defined by SAE International to classify the extent to which the vehicle's automated systems handle the driving task. (Image adapted from SAE International, 2019.)

Thus, prior to the 2021 revision, only Level–2 vehicles were allowed for commercial use. So, the step from Level–2 to Level–3 is a critical one because it transfers the responsibility—and with it, the liability—from the driver to the car manufacturer.[3]

Let us draw a short interim conclusion here: Level–0 and Level–1 can be skipped in our context. Level–0 means no automation, and Level–1 indicates the presence of one electronic support system, e.g. cruise control. From Level–2 on, things get more interesting. Level–2 means that two or more assistance systems interact, e.g. adaptive cruise control and automated lane keeping. But—and this is another source of confusion—SAE leaves open how sophisticated these systems are, which triggered the industry to invent terms like Level–2+ and Level–2++ to indicate hands-free operations. Since 2021, SAE and ISO have differentiated between *Driver Support Systems* (Level–1, Level–2) and Automated Driving Systems or ADS (Level–3 to

Level–5). Level–2 is also often referred to as the Advanced Driver Assistance System (ADAS). Tesla's so-called *Autopilot* or *Full-Self-Driving* (FSD) features are—despite the name—actually Level–2 systems.[4] To make this more transparent, Tesla changed the name to FSD Supervised in 2024.

To differentiate Levels [3–5] even more sharply, SAE introduced a new concept in its last revision in 2018: The *operational design domain.* If an ODD applies, the vehicle is restricted in its operation in a certain way.

The SAE terminology is precise but often criticized for not being product-oriented and consumer-friendly. Consequently, Mobileye proposed a new taxonomy with four levels: *eyes-on/hands-on, eyes-on/hands-off, eyes-off/hands-off, and no driver (robotaxi).* [320] While this taxonomy is more straightforward and more descriptive, it obscures crucial differences, particularly the varying ODD implications of each level.

In 2023, Mercedes-Benz became the first to *homologate* a Level–3 system, meaning it received official road approval. This system allows the driver, in a *clearly defined situation*—its operational design domain—to shift their attention away from traffic and transfer driving responsibility to the system (Eyes-off/Hands-off). A defined situation might include a structurally delimited *median strip* (a physical barrier separating opposing traffic lanes), a maximum speed of 40 mph (60 km/h) and since December 2024, even 60 mph (95 km/h) in the right lane and the presence of a vehicle ahead on a specifically whitelisted, pre-approved road segment.[5] Such whitelisting is centrally managed by an operator based on current information such as road conditions—e.g. no roadworks on this stretch, clear road marking, etc.—weather, or other relevant factors.

Currently, these Level–3 features function as an extended congestion pilot for highways. Plans to raise the speed limit to 75mph (120km/h), potentially upgrading it to a highway pilot, are underway. Still, they are much more difficult as the processing requirements increase exponentially with speed. Plus, increasing the speed limit introduces significant liability issues, making it uncertain when or how this will be implemented.

The DARPA Challenges demonstrated that task-based (narrow) AI control systems were far more effective and efficient than systems attempting to generalize across all possible driving scenarios. However, to achieve full Level–5 automation—driving anywhere, anytime, as a human can—the system *would* need to generalize in this way. It is no coincidence that, because generalised AI does not yet exist and may never exist, Level–5 automation also remains unrealized.

12.2 JOHN MCCARTHY'S 'AUTOMATIC CHAUFFEUR' ARRIVES

People such as William Red Whittaker and Sebastian Thrun were highly enthusiastic about deploying their DARPA-refined technologies into real-world applications and keen to engage in more entrepreneurial ventures. Following his involvement in the second DARPA Grand Challenge, Thrun sought ways to commercially exploit visual data, including footage of desert terrain later repurposed as training material for subsequent DARPA challenges.

In 2006, Thrun began developing a tool designed to create panoramic views from this data. At the same time, Google was pursuing a similar concept with its *Street View* platform. As a result, Thrun, a professor at Stanford University, came into contact with Larry Page, the co-founder of Google. Page showed considerable interest in Thrun's technology, particularly as a means of expanding Google's influence into the physical world, or at least its digital representation. Consequently, Google acquired Thrun's company, VuTool, and brought the Stanford professor on board.

Thrun subsequently hired Anthony Levandowski, another veteran of the DARPA Challenges, who gained notoriety for his attempt to send an autonomous motorcycle, known as GhostRider, to the track of the first DARPA challenge. This robot caused a great deal of excitement in the media, which was increasingly fascinated by autonomous vehicles, particularly motorcycles, with reporters dubbing Levandowski their 'media darling' and 'unofficial DARPA pet'—despite the GhostRider's challenge spectacularly failing when it ingloriously toppled over at the start.

In the second DARPA challenge, Levandowski worked for Thrun as part of the Stanford team. Not a man to limit himself to a single role, Levandowski also collaborated with Thrun on Google's Street View project, acted as a sales representative for Velodyne, and even founded his own company, *510 Systems*.

510 Systems developed software capable of merging data *in real time* from GPS, lidar, and cameras into a cohesive 3D map of physical spaces, which was termed the 'continuous representation of reality'. The technology became a key component in mapping efforts for major companies, including Navteq, Microsoft, and Google, which were among its largest customers.

In 2008, Levandowski founded another company, *Anthony's Robots*, which focused on building a self-driving car. Indeed, Levandowski made history by becoming the first person to bring an AV onto busy public roads during an episode of the TV series *Prototype This*.

Google had just announced its version of the DARPA Challenges, the Lunar X Prize. Following DARPA's *Desert and Urban Challenges*, this competition aimed to send robots to the moon, drawing Whittaker once again to research lunar robotics. A clear theme was emerging: DARPA's role as a hub for AV talent was being supplanted by Google.

Larry Page had been interested in autonomous vehicles long before attending the DARPA Challenges. He had even considered writing his thesis on the technology, but, following his advisor's suggestion, chose *search* as his topic instead—with well-known consequences. A decade after co-founding Google with Sergey Brin, Page found himself once again intrigued by the world beyond search engines. Google's massive global mapping project, Street View, launched the company into the mapping industry, a field previously dominated by the Navteq and TeleAtlas duopoly, requiring immense resources and effort.

The logical next step was to enable computer-controlled vehicles to drive on these newly mapped roads. With Sebastian Thrun and Anthony Levandowski—both leading DARPA veterans—already under contract, Page was well-positioned to explore this frontier. Although Thrun harboured some doubts about the feasibility of autonomous mobility in an open-world environment, which posed challenges far beyond the DARPA trials, he ultimately could not provide a technical reason why it shouldn't be possible, even after careful consideration when explicitly asked by Page to name just one.

Autonomous transportation has additional potential for a lucrative business. Capturing even a fraction of the transportation market could generate massive profits. According to Page, they could create a company as large as Google solely based on this technology. Despite the uncertainty of its success, with only a one-in-ten chance of working, Page believed it was a worthwhile investment of resources, both in terms of time and money.

Thrun decided to pursue the challenge and knew exactly who he wanted on his team: An all-star group made up of members from CMU and Stanford who had dominated the DARPA Challenges. If anyone could complete what DARPA had started and integrate all the lessons learned along the way, it would be this collective of people.

From Stanford, the team included Mike Montemerlo, Hendrik Dahlkamp, Dirk Haehnel, and Anthony Levandowski. Chris Urmson and Bryan Salesky represented CMU. Although Salesky initially stayed at Caterpillar, where he worked with Urmson on autonomous mining vehicles, he later joined the team. Dave Ferguson, a member of CMU's Urban Challenge team and the eventual founder of Nuro, joined at a later time.

Thrun and Urmson were finally united in a group where funding was no obstacle. By 2008, Google was already making a profit of $4.2 billion. The team gradually brought in additional key members from DARPA squads, including Dmitri Golgov, a crucial figure from Stanford's Urban Challenge team. Under Google's banner, the world's best young engineering talent assembled. Though they had the skills to bring the Urban Challenge into the real world, the project remained secret for two more years. Instead of calling it an autonomous vehicle, Google used the codename Project Chauffeur—likely a nod to McCarthy's *automated chauffeur*—and started by licensing the Stanford team's code from the 2007 DARPA Urban Challenge.

The team did not start from scratch, as they had a strong foundation of DARPA technology. Rather than designing a new robot, they improved upon an existing one—following a standard blueprint. Engineers were assigned specific areas, such as perception and planning, allowing them to collaborate efficiently. The robot followed the established DARPA formula, featuring a drive-by-wire system that executed computer commands. For environmental perception, it used radar, cameras, and a roof-mounted spinning lidar system developed by Dave Hall.

The setup was familiar: Thrun acted as the lead while Urmson managed day-to-day operations. The small team reported directly to Larry Page. They began by retrofitting Toyota Priuses, and in spring 2009, they quietly brought them onto the streets. Operating under *Stealth Mode* and navigating a legal grey area, as there were no laws explicitly prohibiting self-driving cars, they deployed two safety operators in each vehicle. When minor accidents occurred, the team discreetly resolved the issues by compensating the other party generously, keeping both the authorities and the press in the dark.

Also, the technical approach mirrored that used in the DARPA Urban Challenge. Mapping the roads was a critical step in preparing for autonomous vehicles. The DARPA Challenges had shown the engineers that lidar was indispensable for self-driving. To create highly accurate maps, lidar-equipped cars were driven to record the precise locations of lane lines, curbs, signs, traffic lights, and other road features with an accuracy of 10 to 15 centimetres. This precision reduced the strain on the car's sensors, enabling them to detect and interpret their surroundings in real time.

With prior knowledge of stop signs, speed limits, and other road features, the car could avoid running red lights or exceeding speed limits. Additionally, these detailed maps helped the car determine its exact position, which is clearly crucial for safe navigation. Unlike GPS, which has a margin of error of up to several metres and requires an optimal line of sight, the lidar-generated maps allowed the car to compare its surroundings to the recorded data, accurately pinpointing its location.

The team had to stretch goals set by Larry Page: (i) to accumulate one hundred thousand miles of autonomous driving on public roads and (ii) to drive ten routes scattered across California, each roughly 100 miles long, resulting in a total of around 1,000 miles, without any human intervention. A challenge internally dubbed 'The Larry 1k'—a challenge that made Tony Tether's DARPA events look like child's play. [54]

In 2011, Sebastian Thrun of the Stanford team launched Google X, the company's *moonshot division*. Google X's biggest bet was the self-driving car. It was also the most ambitious endeavour. Google—now Alphabet Inc.—was already a multibillion-dollar company at the time, but it was highly dependent on its core product, Google Search. This is why Google X—or now just X, the moonshot factory—was intended to be the breeding ground for the next big thing. X's mission was to make the world a better place, and what would be better suited than autonomous driving? And who would have been better equipped to develop this for Google than Sebastian Thrun? The project at Google X was first publicly initiated in 2010 via a post from Sebastian Thrun entitled 'Chauffeur'. It was a well-resourced project, as over six years, Google's spending for Project Chauffeur was in excess of $1.1 billion. [267]

Now that the self-driving program was public, Google needed to leave the legal grey area. Nevada was interested in spearheading the development and passed a bill in 2012 allowing self-driving cars under strict regulations. The car needed to be equipped with a black box recorder for sensor logging, two certified safety drivers, and a significant surety bond. Finally, each vehicle would need to pass a driving test, demonstrating its ability to drive on different types of streets, merge onto and exit the highway, and enter a school zone, adjusting its speed accordingly.

On May 1st, 2012, Urmson and Levandowski passed the test using one of their Toyota vehicles. Google had the first *legally sanctioned* safe-driving car in the world. Closely after Nevada, California strove to catch up. It had stricter rules, including annual disengagement reports, but Google preferred to operate in its home state. Other states followed suit and implemented similar rules. By the end of the decade, autonomous test driving was allowed in half of the states of the USA.

Google was by far the best company operating in the space, but the public attention created interest from others. A gig-economy-based taxi company called Uber, founded in 2010, developed a growing interest in AV technology. In 2013, UberCabs really took off. Larry Page was interested enough to invest in the startup. Uber had the business model, and Google had the technology; it appeared to be the perfect match. The business rationale for Uber was clear: The most expensive cost factor for each ride was the driver, pocketing about three-quarters of each driving fare. Thus, if a computer could replace the driver, Google would finally have a route by which to capitalize on its investments. Page invested $ 258 million into Uber and ordered a Google executive onto the startup's board.

However, instead of proceeding with the original plan and providing the software to Uber, Google began to develop its own car. Using a tier-one supplier to assemble it, the asset was entirely under the control of the internet company. It was dubbed *Firefly*, but because of its animal-like face, the two-seater vehicle was internally referred to as the 'Koala car'. The car was supposed to have no pedals and steering wheel, but that proved premature. Initially, the engineers hacked into a flat disc to steer the car; later, this was replaced by a proper joystick-like handle. Its speed limit was set to 25 mph.

Despite some teething problems, the Koala continued to improve, and with it, Urmson's confidence grew to challenge Uber in its own field. But once Uber's Travis Kalanick got wind of this, he reacted with a fierce counterattack. Since he saw autonomous driving not only as an opportunity for Uber but also as a deadly threat, the reaction had to be swift and decisive. Hence, Travis began scouting to build a new research group that could rival Thrun's DARPA all-star team.

Lewandowski, finding himself locked in an increasingly bitter rivalry with Urmson, also flirted with joining Uber. By now, Google's momentum had begun to slow. There were hiring freezes and increasing corporate demands. Sebastian Thrun left Google in 2014, and Larry Page installed John Krafcik, an auto-industry executive, who demoted Chris Urmson to technical lead. The new entity was rebranded as *Waymo* and became its own company. Two years later, Urmson also quit to fund his own self-driving car startup called *Aurora*—where he teamed up with Sterling Anderson, Tesla's former Autopilot director. [252]

At this juncture, Uber hatched a secret plan for Levandowski to leave Google and build his own company, which Uber could subsequently acquire. And that's what they did. Half of the roughly thirty employees of the company—called *Otto Trucking*—came from Google. Uber acquired the venture for $680 million. The ride-hailing startup hired aggressively to catch up with Google's years-long lead. In so doing, they targeted engineers that Google would have liked to hire but couldn't due to a hiring freeze caused by the increasingly complex corporate processes in the multibillion-dollar Alphabet group. The agile and decisive start-up team, on the other hand, quickly grew to more than 300 people, and the formerly intertwined companies became bitter rivals. The historic rivalry between Stanford University and Carnegie Mellon had morphed into a new feud between Uber and Google.

Thus, Google was no longer alone in the market and faced direct competition. The rivalry resulted in a lawsuit in which Waymo accused Uber of stealing intellectual property, most notably the lidar design. Uber fired Levandowski, who was charged with stealing classified information—more than 14,000 technical documents—from an internal Google server. He was later sentenced to 18 months in prison. Waymo and Uber settled the conflict for a 0.34% share of Uber's equity and a commitment not to use any of Waymo's hardware or software design.

By that time, autonomous driving was everywhere. In the middle of the decade, it had become a common strategic goal. As Alex Davies observed, [77], the early DARPA Challenges sparked a revolution in the field of autonomous driving. The once-sceptical voices about the potential of self-driving cars had now been silenced, and the focus had shifted to determining who would benefit the most from this technology and who would bear the consequences. Almost all major automakers and technology companies joined the race to develop autonomous driving systems. Startups were emerging to seek funding to create software or fill market gaps. Governments were considering enacting regulations and laws to govern this new field of technology.

The world of autonomous driving has evolved into a thriving and rapidly growing ecosystem, with numerous major players involved. General Motors acquired Cruise, Ford took over Argo AI, Intel bought Mobileye, an Israeli company founded by another DARPA veteran, Amnon Shashua,[6] and startups such as Nuro and Zoox emerged with their specialized technologies. Even ride-hailing companies like Lyft and Zipcar were exploring the potential of autonomous driving. Also, Apple launched *Project Titan*, a secret venture aimed at building a fully autonomous car. When rumours about this first project began to circulate, an Apple executive remarked, 'Well, the car is the ultimate mobile device, isn't it?' [280]

This flurry of innovation and the run to driverless cars followed an underlying script, and Larry Burns outlined the narrative. He saw personal cars as a waste of space and energy, especially in larger cities. After his retirement, he published a book called 'Reinventing the Automobile: Personal Urban Mobility for the 21st Century'. Here, Burns presented an influential narrative of a technical development that should lead to fewer CO_2 emissions, less parking space consumption, and better asset utilization.

If a car is only used for one hour a day on average, it sits unused for the other 23 hours. Therefore, the cars of the future should be shared rather than owned, powered by clean energy, able to communicate with each other and make more optimal use of space. Sharing, in particular, could be boosted by AVs: If a vehicle can drive autonomously for nearly 24 hours and pick up passengers, the price of a mile driven would be a fraction of today's cost, rendering car ownership a luxury akin to owning a personal horse. And cities would become more people-friendly. In a text titled 'The Third Transportation Revolution', Lyft co-founder John Zimmer captured this spirit well:

> 'Next time you walk outside, pay really close attention to the space around you. Look at how much land is devoted to cars and nothing else. How much space do parked cars take up lining both sides of the street, and how much of our cities go unused, covered by parking lots? It becomes obvious we've built our communities entirely around cars. And for the most part, we've built them for cars that

> aren't even moving. The average vehicle is used only 4% of the time and parked the other 96%. Most of us have grown up in cities built around the automobile, but imagine for a minute what our world could look like if we found a way to take most of these cars off the road. It would be a world with less traffic and less pollution. A world where we need less parking, where streets can be narrowed and sidewalks widened. It's a world where we can construct new housing and small businesses on parking lots across the country or turn them into green spaces and parks. That's a world built around people, not cars. All of this is possible. In fact, as we continue into our new century, I believe we are on the cusp of nothing short of a transportation revolution, one that will shape the future of our communities.' [392]

Having said that, Zimmer (*ibid*) focused on the solution that technology could provide for this problem:

> 'Technology has redefined entire industries around a simple reality: You no longer need to own a product to enjoy its benefits. With Netflix and streaming services, DVD ownership became obsolete. Spotify has made it unnecessary to own CDs and MP3s. Eventually, we'll look at owning a car in much the same way.'

Concluding that (*ibid*):

> '[. . .] once autonomous networks provide better service at a lower cost, our country will pass a tipping point. And by 2025, owning a car will go the way of the DVD. Until then, over the next five to 10 years, there will be both driver and driverless cars on the road, which we call a hybrid network.'

The reference to the 'hybrid network' highlights an alternative approach to full mobility, adopted by another player in the Silicon Valley tech startup scene, Tesla, which we have not yet discussed: Privately owned robotaxis. Tesla's CEO, Elon Musk, presented the counterargument, asserting that the transition to autonomous vehicles (AVs) would not occur through shared fleets but rather through a network of privately owned AVs, which owners could rent out to others when not in use.

Tesla was a hardware manufacturer. The electric drive pioneer built vehicles that could travel hundreds of miles between battery charging stops. At the same time, the cars were sporty and designed for high acceleration. Tesla had the classic ownership-based business model of automakers, but at the same time, it was a tech startup with a big focus on software and connectivity. Wireless software updates not only made the cars more intelligent and more efficient, but they could even provide a new set of features. This, Musk argued, did not make the cars less valuable over time but could maintain or even increase their value. In October 2015, one of those over-the-air upgrades enabled the vehicles to drive themselves.

About a year earlier, Tesla had started equipping all its vehicles—regardless of whether the buyer configured them or not—with the necessary hardware, such as cameras to detect lanes and radar to maintain a safe distance from other cars. Just a fee-based software upgrade was needed to enable these preinstalled components to realize their intended function.

Tesla had a very different strategy from Google/Waymo. While Google technically pursued the DARPA approach, maps, and lidars, Tesla took a lightweight approach limited to cameras and radar, and entirely centred around AI. Their business cases were also quite different: Google, like Uber, relied on vehicle sharing, while Tesla relied on vehicle ownership.

Given this, it is interesting to note that in 2013, Google was negotiating a purchase of Tesla. Tesla was on the verge of bankruptcy at the time, and Musk had contacted his friend Larry Page for help. As The Guardian newspaper outlined, a deal was written. Musk demanded that Google buy Tesla for $6 billion, with a $5 billion sum for factory expansions, with guarantees that Google wouldn't simply break up the company and that Musk could run it for eight years or until it produced its third-generation car. [115]

At that time, Larry Page was still searching for a viable business model and monetization strategy. However, before we examine the different monetisation approaches, we need to revisit the technologies that underpin these fantasies.

12.3 AI AND AUTONOMOUS DRIVING: MACHINE LEARNING OR RULES?

The process of autonomous driving can be organized into four stages: Sensing, perception, planning, and control. Sensing and control use sensors and actuators, respectively, to gather data about the *surroundings* and to *control* the AV. However, as we have seen, perception and localisation require *machine learning* to adapt to changing environments in real time. The perception module creates a model from the available environmental data. Typically, the planning module uses rules to calculate vehicle paths and conduct driving manoeuvres, such as stopping at a traffic sign.

Sensors, such as cameras and lidar, and pose devices, including GPS and IMUs, gather data about a vehicle's surroundings and position. In the perception stage, AI integrates all sensor data in an attempt to *understand* the vehicle's surroundings. Since Sebastian Thrun, artificial neural networks have been successfully deployed for elements of this task. However, path planning and vehicle control still primarily rely on rule-based algorithms and classical control models, at least in higher levels of automation.

In Level–2 ADAS systems, rules are becoming increasingly replaced by end-to-end neural networks, and the respective vendors aspire to extend this approach to *Automated Driving Systems*, e.g. Tesla with its robotaxi announcements in 2024 based on *FSD Unsupervised.*

Path planning is typically implemented through a combination of global and local planning systems. Once defined, control algorithms calculate parameters that serve as control inputs for the vehicle's actuators. Until 2010, engineers predominantly resorted to rule-based AI in their decision-making and high-level control. This requires codified knowledge to be symbolically represented as comprehensively and clearly as possible. As computer scientist and Berkeley professor Stuart J. Russell highlights, AVs need rules:

> 'Many driving situations need classical rules for an AI to be successful. For example, if you're in the middle lane and you want to change lanes to the right, and someone is trying to pass you on the inside, then you should wait for them to go by first before you pull over. For road situations that require a lookahead, because no satisfactory rule is available, it may be necessary to imagine various actions that the car could take as well as the various actions that other cars might take, and then decide if those outcomes are good or bad.' [104]

However, an open question is whether these rules need to be encoded or can be learned by the algorithms directly from driving data. In the third DARPA challenge, we observed issues arising from over-reliance on a manually programmed, rule-based approach, as exemplified by the Penn Team's *Little Ben* vehicle. Here, knowledge about driving and the world was embedded in a complex rule set and knowledge base that required careful engineering and curation by hand. That presented a fundamental challenge because, when operating in an open, unconstrained world, such a rule set and knowledge base is, in principle, never complete.

Nonetheless, early versions of self-driving cars continued to rely on rule-based approaches. For example, Google's Koala-car, Firefly, employed specialized hardware systems to detect other vehicles, lane markers, obstacles, and pedestrians. The vision systems would pass this information on to a controller, which applied logical rules to determine what the car should do.

However, as Google continued to test cars on public roads, it found itself adding new rules every day. In an open world, this rule-based approach can never be complete. Every day, new situations arise, and new rules are introduced. An approach that is not feasible in the long run—omitting a particular rule for some situations—could prove a matter of life and death on the road. John Hayes, CEO of the former AV 2.0 Startup Ghost Autonomy, called this 'the robotics problem of AV':

> 'It is clear that robotics, the approach that has dominated the past decade of AV development, has proven inadequate to solving the massive complexity of real-world driving. [. . .] Robotics is fundamentally defined by constraints, which are painstakingly programmed by engineers. A look into the classic autonomous vehicle software stack reveals a long list of rules about what a computer cannot do: HD maps tell it where it can't drive, object recognition tells it what it can't hit, and pre-programmed rules of the road tell it how it can't behave. What's left—the drivable surface minus impediments—is the path forward. But what worked well on an obstacle course has proven inadequate for the complexities of real-world driving.' [134]

The *robotic approach* fails when conflicting constraints arise. Such situations require a human, like the Google engineers in the Koala car, to determine which rule takes precedence in which instance. It is a complex, labour-intensive and error-prone process that necessarily only works until another exception is discovered. Even worse, it forces an additional layer of complexity. And another layer of human assumptions, which makes it subject to additional human error or bias. It also builds in conflict, as, by definition, a rule that had been deemed necessary may need to be deemphasized in certain situations. This renders the initial assumptions more challenging to change; with every exception built upon every rule, they become increasingly difficult to adjust.

As the AV executes code, there is no generalizable way to filter or prioritize incoming data, and the intricate balance of rules, exceptions, and exceptions to those exceptions creates a significant risk of system failure due to information overload. As John Hayes explains:

> 'The number of edge cases increases by the square of the number of rules. If you have 100 rules in your model, and each rule can conflict with every other rule, you have 10,000 possible edge cases. The more complex the driving scenario, the more rules are required to drive safely—and the number of edge cases explodes. A complex situation might require 1,000 rules, which would produce 1,000,000 possible edge cases. Resolving these by hand is nearly impossible.' [134]

A decade after the DARPA Urban Challenge, the ability to deal with unexpected situations remains the biggest challenge for AVs, and a solution to this problem will only begin to emerge with the development of more advanced forms of machine learning: systems that can automatically learn and adapt to new contexts as they arise. If possible, a shift away from logic and handcrafted rules towards adaptive machine learning appears to be a promising way to bring about significant progress in the development of self-driving cars.

The advent of *end-to-end* learning (see Section 16) and multi-modal neural machine learning (see Section 15.5) promised a solution by replacing rules entirely with machine learning. Early adopters, such as Ghost Autonomy, recognised this opportunity as early as 2020 and received financial backing from OpenAI to pursue the idea. The most significant advantage of machine learning is that the algorithm can identify patterns in the world that people fail to recognize; as Chris Anderson's mantra suggests, data speaks for itself. Thus, start-ups like Ghost treated driving as a machine-learning problem, not a rule-based problem.

The key element in driving a real-world scenario is determining what information is essential and what can be ignored. Humans are good at filtering out noise, not at memorising many rules and exceptions. Again, in the words of Hayes:

> 'The core challenge is that there is a lot of noise on the road. The total data input at a given moment on the road is massive, but only a small fraction of it is important in making your next driving decision. Consider the freeway—despite a constant stream of cars whizzing past you in both directions, moment by moment, almost all of your decisions are based on a few lane markers and the distance between you and the car in front of you. But when a car swerves towards you

from two lanes over, what was an irrelevant input moments ago becomes central to your split-second decision. Determining what data input is important at which moment is the heart of the task.' [134]

Similar to Chris Anderson's *end of theory*, the machine learning approach does not start with theoretical assumptions about driving. Instead, it begins with examples of how people actually handle driving tasks in the real world, utilising deep learning and harnessing the *unreasonable effectiveness of data*. Capturing what people see in a particular situation and what they decide to do next serves as the ground truth for the learning model. Machine learning techniques then identify which features in the environment actually influenced each driving decision. As Hayes (*ibid*) explains:

> 'This process is highly iterative, testing a broad selection of potential inputs and measuring model performance against more real-world data. You are essentially developing a driving model in reverse, starting with the right answer, and then using math to discover how people get to that answer.'

Although Tesla vehicles currently remain firmly pegged at Level–2 autonomy, in 2024, Tesla also announced that it would replace planning algorithms with artificial neural networks within its end-to-end architecture. Currently, companies operating at higher levels (Level–3 and Level–4) continue to use rule-based algorithms for planning and control. These algorithms remain necessary—at least to date—to ensure key safety features such as determinism, real-time capability, and transparency. Nevertheless, Musk seems confident that the *unreasonable effectiveness of data* can be applied to solve rule and logic-based problems. To address the challenges, Tesla will rely on big data, statistics, and scalable computation. Interestingly, this is an analogous revolutionary approach to the one introduced by Sebastian Thrun in the second DARPA challenge, a method that led to a decisive breakthrough. Now, those former DARPA iconoclasts had become staid. And Elon Musk aims to advance this development even further. On April 9, 2023, he posted on X:

> 'We need to predict what pedestrians will do based on their behaviour, including limb angle & direction of sight. FSD currently sees all pedestrians as cuboids, so it is overly cautious. Also, diffusion seems to be more compute-efficient than transformers for vision.' [255]

Traditional convolutional neural networks process images hierarchically, whereas vision transformers treat images as sequences of patches. This approach enables vision transformers to capture global dependencies and model long-range pixel-level interactions. Their scalability also allows them to handle larger input image sizes effectively. Diffusion models, which are even more scalable, excel in tasks with highly variable input data. This makes them particularly suited for complex predictions, such as analysing large video datasets from real-world driving scenarios. In subsequent sections, we will delve deeper into these models.

Notes

1. Since the third DARPA Grand Challenge, autonomous driving technology has made considerable progress, though primarily in structured and predictable environments. In more complex settings, such as off-road military terrain, autonomous vehicles (AVs) have struggled to match the performance of human-driven cars, mainly due to limitations in their autonomy algorithms and software. DARPA's RACER programme seeks to overcome this challenge by developing technologies that enable unmanned ground vehicles to navigate unstructured off-road terrain, pushing their mechanical limits while matching or exceeding the speeds and efficiencies of human-driven cars. The programme aims to create off-road, unmanned vehicles capable of reacting like humans, building on advances in on-road AV technology and adapting them for unpredictable environments.

2. SAE recognizes this as well. However, the latest revision of the J3016 definition overview describes Level–5 as: '*this feature can drive the vehicle everywhere in all conditions*', [294] the 2014 version was more restrictive: '*(Level–5 is) the full-time performance of an automated driving system of all aspects of the dynamic driving tasks under all roadway and environmental conditions that a human driver can manage.*' [291] This introduced a fundamental limitation: Only under conditions that a human driver can manage. It is unclear why this clarification is missing in the updated public version. The 2018 revision states: 'the feature can drive the vehicle under all conditions' and: '*(Level–5 is) the same as Level–4, but the feature can drive everywhere and under all conditions.*' [293] This could imply that the system can operate in conditions beyond human capability, which is incorrect, as clarified in the full version restricted to SAE members. The full 35-page document distinguishes between unconditional and operational design domain (ODD)-specific operations: '*Unconditional/not ODD-specific means the ADS [Autonomous Driving System] can operate the vehicle under all driver-manageable road conditions within its region of the world.*' [292]

3. In practice, this will be a difficult question for courts to decide, as car manufacturers will take on product liability (which they have anyway) in case of Level–3 driving, but will not accept tort liability for 'negligent homicide' or similar charges, even if the driver fully adhered to the duties of the owner's manual. [188]

4. In October 2024, Tesla announced that it was working on a system for higher Levels than Level–2, called *FSD Unsupervised*. It was announced for 2027 at the earliest. However, whether this is feasible without HD maps, operational design domain restriction and lidars remains an open question. To distinguish the Level–4 system from the current FSD Level–2 system, the ADAS version is now called FSD Supervised.

5. In September 2024, Mercedes announced it would be working with authorities on road clearance for a Level–3 system with a speed of up to 55.9 mph (90 km/h).

6. In the 2005 DARPA Grand Challenge Amnon Shashua was co-lead of the Golem Group/UCLA team that scored 14th. [226]

Section V

Attention, Diffusion, and End-to-end Learning

The invention of neural architectures capable of *attention* and the subsequent recent rise and rise of the transformer architecture rapidly led to the development of large language models and the now ubiquitous ChatGPT chatbot, with LLMs fast emerging as the dominant technology whenever AI is discussed in the popular press.

With the development and subsequent hegemony of LLMs, it seems that sixty years after the Dartmouth conference, the view of AI as synonymous with one technology, or overarching framework, may have finally arrived; in the words of AI researcher Francois Chollet of Google:

> 'With the resurgence of machine learning in the 1980s, its rise to intellectual dominance in the 2000s, and its peak as an intellectual quasi-monopoly in AI in the late 2010s via deep learning, a connectionist-inspired Tabula Rasa is increasingly becoming the dominant philosophical framework in which AI research is taking place.' [61]

And it is with the development of the transformer model and end-to-end neural control systems that we position the current *third wave of vehicular autonomy*.

Note for readers: *In 'Driving Intelligence', we've aimed to keep the technical and mathematical aspects of AI relatively light. However, as transformers and diffusion models lie at the heart of current practice, we've chosen to explore these topics in more detail.*[1] *While we've worked to make these sections as approachable as possible, readers who prefer not to delve deeply into the technicalities can feel free to skim or skip this material.*

13 Recurrent Sequence-To-Sequence Learning

'Sequence-to-sequence' (seq-to-seq) learning defines a neural network architecture designed to transform one sequence of data into another. Seq-to-seq was initially used in applications such as machine translation, text summarization, speech recognition, etc., where there is a structured relationship between input and output sequences, albeit these may differ in sequence length or type.

A seq-to-seq model typically consists of two main components: an *encoder* and a *decoder*. The encoder processes the input sequentially, generating *hidden states* at each step that reflects the model's understanding of the input up to that point. The final hidden state becomes the *context vector*. The decoder operates similarly, predicting the next token in the sequence based on the context vector, its hidden state, and the previous output token until an end-of-sequence token is generated, a process which can lead to prediction errors compounding. These predictions are made as probabilities over the possible tokens, typically using a *softmax* function.[2]

Typically, seq-to-seq models are trained by minimizing the difference between predicted and target sequences via a *cross-entropy loss* function.[3] During training, a technique called *teacher forcing* is commonly employed, where the actual token from the target sequence is passed to the decoder as input at each step.

Seq-to-seq models were first constructed using classical recurrent neural networks (RNNs) and the more complex Long Short-Term Memory (LSTM) recurrent networks. These have performed well on some seq-to-seq tasks; however, they struggle to process long sequences accurately. The introduction of *attention mechanisms* and *transformer models* has significantly improved seq-to-seq performance.

13.1 STATIC WORD-EMBEDDINGS

In seq-to-seq *text processing* applications, an input sequence of *words* is first transformed into a sequence of *static embeddings—vectors*, whereby words/tokens with similar meanings are projected into similar regions of the vector space. For example, *word2vec* is a group of related models that are often used to produce *word embeddings*. These models are shallow, two-layer neural networks that are trained to map words to vectors that define their linguistic context(s). Thus, word2vec takes a large corpus of text as its input, which it maps to a vector space, typically of several hundred dimensions, with each unique word in the corpus assigned a corresponding vector in the space. The transformer, as originally described, [358] used 512 embedding dimensions ($d_{embed} = 512$), but in practice, any reasonable value can be used.

DOI: 10.1201/9781003561460-18

Thus, a proxy for the *semantic similarity* between two words can be established by calculating the similarity between the two vectors, as defined by the *cosine similarity* of their embeddings, for example.[4]

13.2 RECURRENT NEURAL NETWORKS

Classical recurrent neural networks (RNNs) are a form of neural network that extends the basic feedforward multilayer perceptron architecture. The critical distinction is that the output of a recurrent neuron is fed back as an additional input at the next time step, alongside the next value in the input sequence. Consequently, information in RNNs flows both forwards and backwards through the network. This *recurrence* is the critical differentiating factor that separates RNNs from classical *feedforward* networks, as the latter do not involve such feedback. Typically, like MLPs, during training, weights and biases are optimized using a gradient descent method, such as backpropagation learning.

Consider a sequence of inputs, denoted as x_t, where t represents the current time step. In natural language processing, the input sequence might represent a sentence, with each x_t corresponding to a vector embedding, a structure representing the word, seriatim (as described in Section 13.1). At the initial time step ($t = 0$), the input to the neuron is x_0. This input vector undergoes standard feedforward processing: it is multiplied by a weight w_x, and a bias term b is added to produce a weighted output. This output is then passed through an activation function. In recurrent neurons, this is usually the sigmoid or hyperbolic tangent (tanh) function. The resulting output, y_0, is subsequently fed back into the neuron at the next time step, together with the following input value, x_1. This process repeats for every value in the input sequence.

The output y_0, encoding information about the first input, is termed the *hidden state*. As t increases, the hidden state captures cumulative information about all preceding inputs in the sequence, i.e. each neuron's output at step t, y_t, depends on the current input, x_t, and the aggregated information from prior inputs encoded in the hidden state. Multiple layers of recurrent neurons are connected to form the complete RNN.

Unfortunately, RNNs suffer from two key drawbacks: (i) the vanishing/exploding gradient problem and (ii) a limited context window. The exploding gradient problem and vanishing gradient problem can be attributed to the recurrent nature of RNNs, where each neuron's output, y_t, is influenced by the output from the previous time step, y_{t-1}. When the current output value is large, such as numerically exceeding 2, subsequent outputs increase progressively, resulting in a compounding effect over multiple iterations. For example, in a sequence of ten inputs, an initially large value of y_0 will lead to progressively larger outputs, ultimately producing an exponentially large result. This is known as the exploding gradient problem and poses a significant challenge during training. Exploding gradients can cause the gradient descent algorithm to take very large steps, preventing convergence to optimal weight and bias values and leading to poor network performance.

In contrast, if the current output value is small, such as less than 1, the subsequent outputs decrease progressively. This effect, known as the vanishing gradient problem, results in diminishing gradient steps that are insufficient to adjust the weights and biases effectively. As a result, the network fails to converge to optimal parameter values, adversely affecting its ability to learn and perform accurately.

Another challenge associated with RNNs is the limited context window, which arises from the way hidden state values encode information. For short sequences, the hidden state, represented by a single floating-point value, can adequately capture information about prior inputs. However, as the sequence length increases, such as in long paragraphs or entire books, the hidden state becomes dominated by recent inputs and loses its dependency on earlier values. This phenomenon, referred to as the limited context window, can be interpreted in natural language processing as the network 'forgetting' earlier words and sentences, thereby reducing its ability to model long-term dependencies effectively.

13.3 LONG SHORT-TERM MEMORY

A Long Short-Term Memory (LSTM) cell, pioneered by Hochreiter and Schmidhuber, is a complex type of recurrent neural network neuron, termed a *cell*, which is designed to process sequential data efficiently. [144] Unlike traditional RNNs, which can struggle with vanishing or exploding gradients, LSTMs can better capture long-term dependencies within sequences.

The key to an LSTM's power lies in its three *gates*: The forget gate, the input gate, and the output gate. The forget gate determines which information from the previous cell state should be discarded. The input gate decides which new data from the current input should be stored in the cell state. The output gate controls the flow of information from the cell state to the output.

Thus, the *forget*, *input*, and *output* gates work together to manage the memory of the cell. At every step, the LSTM examines the current input and the previous hidden state, determines what to forget, what to remember, and what to output, and then updates its memory and hidden state for the next time step.

As an analogy to LSTM operation, imagine a student preparing for an exam over several weeks. As they study, they periodically review their notes and decide which topics they no longer need to focus on, such as concepts they have already mastered. This process resembles the forget gate in an LSTM, which discards unnecessary information. Simultaneously, the student identifies new material to learn from recent lectures and incorporates it into their study routine, much like the input gate that adds relevant information to the memory.

Over time, the student's overall understanding of the subject evolves, retaining important topics while discarding less relevant details. This ongoing refinement mirrors the role of the cell state in an LSTM, which acts as the long-term memory. When the day of the exam arrives, the student carefully selects specific knowledge to recall and use in their answers, similar to the output gate, which determines what information is used at each step.

Like the student, LSTM balances forgetting, remembering, and utilizing information to process sequences effectively. By selectively remembering and forgetting information, LSTM cells can effectively process complex sequential patterns. This makes them a helpful tool for tasks such as language modelling, machine translation, and speech recognition.

Notes

1. We have delved into these topics because a deep understanding is essential for independently assessing the current state of technological progress and aligning with our broader arguments about AI. This knowledge is not only relevant to scientists and technical experts but also critical for policymakers and financial stakeholders, who must address the wider implications of AI and AGI. The technical content is crafted to provide the necessary tools to analyse and navigate the core concepts driving current AI discussions.

2. The softmax function is a mathematical operation that turns a list of numbers into probabilities. It ensures the numbers are scaled such that they all add up to 1.

3. Cross-entropy loss function is a widely used loss function in machine learning, especially for classification tasks. It measures the difference between the predicted probability distribution (derived from the model) and the proper distribution (based on actual labels). This loss function measures the degree to which the expected probability aligns with the actual class labels.

4. For two vectors x_1 and x_2, $\text{cosine\_similarity}(x_1, x_2) = \frac{x_1 \cdot x_2}{\|x_1\| \|x_2\|}$.

14 Attention Is All You Need

14.1 ATTENTION MECHANISMS

In their groundbreaking and highly influential paper—*over 148,500 citations at the time of writing*—the transformer architecture was introduced by Ashish Vaswani et al. at NIPS 2017 as a new deep learning alternative to traditional recurrent neural networks for sequence-to-sequence learning tasks such as language translation, speech recognition, and text generation.[1]

Subsequently, in its various manifestations, the transformer has become the standard approach for almost every state-of-the-art model in natural language processing, including *ChatGPT*, *Gemini* and *Claude*, and more recently, as *vision* transformers, have become central to the end-to-end control pipelines now being deployed in many autonomous vehicles, such as those from Tesla. Due to its significance in current approaches to autonomous driving, we will undertake a detailed examination of the transformer architecture and its key components. We will greatly benefit from this knowledge, specifically in the Red Book, as we analyse potential ways to move forward.

Attention mechanisms mirror aspects of human perception. People, unlike convolutional neural networks, do not laboriously scan every pixel in their field of view with the same emphasis. Instead, humans focus on *specific elements* of a scene by selectively paying *attention*. Hence, a core intuition is that, likewise, neural networks should learn not only by analysing everything but by understanding where to *focus attention* in a given *context*.

By way of analogy, we can consider a transformer akin to a group of people in a room having a discussion. Each person represents a word in a sentence. Instead of only listening to the person next to them—*as in traditional recurrent neural network models*—everyone in the room listens to everyone else and weighs up how important each other's words appear. Some people's comments might be more relevant, and the transformer pays more attention to them. Additionally, even though everyone can talk at the same time (allowing parallel processing), they still remember who spoke first or last using a unique identifier (a *positional encoding*), so the order of conversation isn't lost. This allows the model to *understand* both the relationships and the entire sequence of the words.

We position the first significant use of ANN attention mechanisms with the 2014 paper 'Neural machine translation by jointly learning to align and translate' by Yoshua Bengio et al. [18]. However, just a little later, in 2014, Ilya Sutskever, Oriol Vinyals and Quoc Le also described an attention mechanism in the context of sequence-to-sequence learning.[2][336]

ANN attention mechanisms are helpful because they enable models to dynamically focus on the most relevant parts of the input data when making predictions or generating representations. For example, in text processing, we need to focus on the

DOI: 10.1201/9781003561460-19

keywords that most contribute to meaning, ignoring less important words like articles or conjunctions; similarly, in image processing, when identifying an object in a cluttered scene, we need to concentrate on *specific regions* rather than processing all the elements of the *entire scene* equally.

This ability to selectively attend to specific aspects of the data enhances performance across a wide range of machine learning tasks, particularly in contexts involving complex or high-dimensional inputs, such as text, images, or multimodal data. Furthermore, as we have seen, classical approaches to seq-to-seq modelling, such as recurrent neural networks, have difficulty modelling long-term dependencies due to vanishing gradients or limited contextual memory, as required in, say, language translation. In contrast, the use of an attention mechanism enables direct connections between distant elements, regardless of their position in the input sequence and effectively allows models to consider large input sequences when making predictions.

14.1.1 DYNAMIC WORD EMBEDDINGS

Word2vec is an example of a *static word-embedding*, which, while widely used, suffers from a significant limitation: the absence of contextual information. This drawback becomes particularly apparent when dealing with words that have multiple meanings, known as *polysemous words*. Such words are encoded with inherently *ambiguous representations* because static embeddings do not incorporate the *context* necessary to distinguish between different *meanings*.

Consider the word ‘bank’ as a classic example of polysemy. In a static embedding model, *bank* would be represented in a vector space in a way that reflects some similarity to words like *money* and *deposit* while also showing some similarity to words like *river* and *nature*. This ambiguity naturally arises because the word ‘bank’ will appear in a variety of contexts throughout the training data. The inability to adapt to these varying contexts highlights the fundamental issue with *static* embeddings: their representations remain fixed and do not adjust to reflect subtle contextual changes.

The Attention block addresses the shortcomings of static embeddings by generating *Context-aware embeddings*. In this approach, fixed word embeddings are enhanced with two additional types of information: *positional information*, which indicates where each word appears in the input text, and *contextual information*, which captures the manner in which words are used. These enhancements are achieved through distinct components within the architecture, specifically *positional encoding* and the *self-attention* blocks.

The positional encoder introduces information about the relative positions of words, enabling the model to interpret the sequence’s word order. The self-attention mechanism, on the other hand, incorporates contextual information by examining the relationships between words across the entire input. Together, these components allow transformers to produce far more nuanced vector representations of words that account for both their position and usage within the input sequence.

By extending beyond the limitations of static embeddings, attention blocks enable models to handle polysemous words effectively, providing representations that are context-specific to the words’ appearances. This capability represents a significant advancement in language modelling, allowing transformer-based models to achieve much deeper sequence-to-sequence processing compared to earlier methods.

14.1.2 POSITIONAL ENCODING

For transformers, before applying the self-attention mechanism, a positional encoding function is applied to the learned embeddings to add information about the order of tokens in the input sequence. These functions must adhere to specific constraints to ensure their effectiveness; they must be *bounded*, produce values constrained within a range, e.g. between 0 and $+1$ or -1 and $+1$, *periodic*, creating repeating patterns that the model can learn to recognize, and *predictable*, enabling the model to understand positional information even for sequence lengths not encountered during training. These constraints ensure that the positional encoder enables words to *attend to* and derive *context* from any other words in the sequence, irrespective of their relative positions. In theory, with sufficient computational power, this design enables words to gather context from every relevant word, even in arbitrarily long input sequences.

14.1.3 SELF-ATTENTION

Transformers are grounded on the following linguistic intuitions: (a) words have many related *senses*; (b) disambiguation is based on *context* (including word order/position); (c) relevant context can be *non-local* and *non-linguistic*; (d) compositional meaning depends on what individual words *mean*, and (e) understanding requires *background knowledge*. These insights informed the design of the *self-attention* mechanism at the heart of transformers. This mechanism enables the transformer to attend to *all* elements of the input sequence *at once* rather than processing the sequence *seriatim*, one word at a time.

Thus, the core intuition behind self-attention is that words derive meaning and context from other words in a sentence, across the paragraph containing the sentence and also, albeit usually to a lesser degree, across the entire linguistic structure being processed. In principle, a sufficiently powerful computer should be able to derive context for any given word by analysing all relevant words within an arbitrarily long input sequence. This relates to what is known as the *context length*—the extent of a sequence from which a model can extract meaningful information. In chatbots like ChatGPT, the context encompasses not only the current input provided by the user but also all prior exchanges within the same conversation, including earlier prompts and responses. However, this is constrained by the *context limit*, which defines the maximum number of tokens the model can process at once. For instance, GPT-3 has a context limit of up to 4,096 tokens. At the same time, the enterprise version of GPT-4 supports much larger sequences (at the time of writing, accommodating up to approximately 128,000 tokens).

Thus, given a sequence of initial word embeddings, self-attention first ascertains the semantic similarity, the *similarity score*, of each word embedding with all others; a proxy for this is the similarity between each of their embedding vectors, and one measure of this is their vector dot product. As Bradney Smith outlines, [328] there are several methods for measuring the similarity between two vectors, broadly classified as *distance-based* and *angle-based* metrics. Distance-based metrics assess similarity by calculating the straight-line distance between vectors, akin to applying Pythagoras's theorem in the embedding space. While this approach is simple, it is computationally expensive in high-dimensional embedding spaces.

Angle-based metrics include *cosine similarity* and *dot product similarity*, both of which define similarity using the cosine of the angle, θ, between two vectors. When vectors are orthogonal (at right angles), $\cos(\theta) = 0$, indicating no similarity. For parallel vectors, $\cos(\theta) = 1$, signifying identical vectors.

However, cosine similarity, which relies solely on the angle between vectors, has two significant drawbacks. First, it ignores vector magnitude, leading to inflated similarity scores for distant vectors that are merely aligned. Second, it involves computing the dot product and then dividing by the product of the magnitudes of the vectors, making it unnecessarily computationally expensive. Consequently, a simple dot product, i.e., $x_1 \cdot x_2 = \sum_{i=1}^{n} x_{1,i} x_{2,i}$, is preferred, albeit to prevent values from getting too large, the dot product is usually scaled by the square root of the dimensionality of the embedding dimensions, d_{embed}.

Next, we need to adjust/move each embedding according to the context defined by the entire sequence of words, with less relevant words causing a slight adjustment and the most pertinent words causing a significant shift. To do this, we first normalize the similarity scores to have a sum of 1, for example, by applying a Softmax function, and then adjust the position of each of the word embeddings by adding the normalized similarity scores of each of the other words to it, so obtaining the new *context-infused* weighted embedding for that word; a process repeated [in parallel] for each of the other words in the sequence. This sequence of weighted embeddings is the output of the self-attention layer.

In the core self-attention mechanism, as described above, the initial word embedding vectors are used three times: first, as the initial embedding vector; second, to calculate the similarity of each word in the sequence with all others, including itself (via the dot product similarity metric); and third, to multiply this similarity by each initial embedding, summing across the entire sequence of initial-embeddings, to obtain the new context-infused embeddings. By analogy with a database query, we can label these three usages of the initial embedding terms: *Query*, *Key*, and *Value*.

Conceived in this way, the self-attention mechanism is a neural architecture that mimics the process of information retrieval, as it measures the similarity between a specific *query* 'q' and each *key* 'k_i', assigning a weight to each key; the output being the weighted sum of all the *values* 'v_i' in the database, i.e. the sequence of words/tokens being processed:

$$\text{Attention}(q,k,v) = \sum_i \text{similarity}(q,k_i) \cdot v_i$$

Thus, attention weights can be calculated by taking the vector embedding of each word in the sequence (the *query*) and calculating its semantic distance from every other word in the sequence (the *keys*) by calculating their scaled dot product. These attention weights are normalized so that they sum to 1.

Then, when we multiply and sum the attention weights with each word in the sequence, we essentially reweigh all words in the sequence relative to each other's *value*. In this way, self-attention serves as a computational proxy for establishing the context of words within a sequence. One of the most exciting facets of the mechanism is that it doesn't rely on word order, word proximity, or sequence length and, in its basic form, does not require learning (iterative weight training).

14.1.4 LEARNING WITH SELF-ATTENTION

So far, we have described self-attention without any form of learned parameters. However, performance can be significantly improved by learning a suitable set of bespoke parameters for the model. There are two ways in which this is achieved in the self-attention block: firstly, learning a bespoke set of initial word embeddings and secondly, additional weights can be introduced in the form of matrices of reals, which are pre-multiplied by their respective inputs to form the *key*, *query*, and *value* matrices (K, Q, and V), where the number of columns in each of these weight matrices is a *hyper-parameter* that can be adjusted based on the model's implementation and use case, albeit they are typically set equal to d_{embed}.[3]

Typically, all weights in the transformer are trained using back-propagation. During each training step, the model takes an input word and attempts to predict the next word in the sequence, a task known as next token prediction (NTP). Initially, predictions are poor, leading to updates of all weights, biases, and the initial word embeddings to minimize the loss function. Training continues iteratively until the model achieves satisfactory performance on the NTP task.

Thus, attention mechanisms weigh the importance of different parts of the input data when making predictions. This ability to focus on relevant information and ignore irrelevant parts is particularly beneficial in tasks where context is crucial. Additionally, the attention mechanism can be computed in parallel, making the architecture highly efficient and scalable. Most importantly, *attention* provides an information retrieval mechanism, a kind of memory, enabling them to deal with larger and more complex problems. This memory allows self-attention to capture long-range dependencies and relationships within the data, which is essential for understanding the context and structure of many tasks.

14.1.5 MULTI-HEAD ATTENTION

The multi-head attention (MHA) mechanism works by effectively performing multiple passes through the input sequence, with each pass attending to different elements of the sequence in parallel, helping the model better understand the *context* of the words in the sequence. Heads differ not in the tokens they attend to but in the features (embedding subspaces) they use to calculate attention.

In the basic self-attention mechanism, as outlined above, a single set of weight matrices (W_Q, W_K, W_V) transforms the token embedding matrix X into *query*, *key*, and *value* matrices (Q, K, V) to compute attention scores and contextual information. MHA extends this by dividing the attention mechanism into H smaller heads, each operating on a subspace of the embedding dimensions, with dimensionality (d_{embed}/H). Each head learns a unique set of weight matrices (W_i^Q, W_i^K, W_i^V), which project the full embedding space (d_{embed}) into a smaller subspace (d_{head}). Working in a reduced subspace encourages each head to explore distinct features or patterns in the input rather than all heads redundantly focusing on the same features.

The MHA design allows each head to specialize in capturing different relationships within the sequence, producing distinct attention outputs. Finally, after computing attention for each head, the outputs are concatenated and transformed back into the original dimensionality d_{model}, via a learned linear projection matrix W^O. This aggregation merges the diverse information captured by each head, enabling the model to synthesize insights from multiple perspectives to create a single output sequence of context-aware embeddings within the original d_{embed} dimensions.

By employing multiple heads, MHA enhances the model's learning capacity, enabling it to capture diverse relational aspects within the data. Different heads can simultaneously focus on varied input characteristics—such as long-range dependencies, grammatical syntax, or short-range interactions—hence producing a more sophisticated sequence understanding.

14.2 THE TRANSFORMER

The transformer architecture is built upon an encoder-decoder structure, similar to the RNN-based translation models discussed earlier (see Section 13). As illustrated in Figure 14.1, from Vaswani's 2017 'Attention is All You Need' paper, [358] the architecture consists of two components: the *encoder* shown on the left of the image, in Vaswani's original description comprising of six encoder blocks, and the *decoder* on the right, comprising of six decoder blocks.[4]

A core innovation of the transformer is the multi-head attention mechanism (see Section 14.1.5), which plays a pivotal role in its success. Attention provides context by determining how different elements of the input and output sequences interact. Self-attention focuses on the relationships among words within the input sequence to derive context. In contrast, *cross-attention* considers interactions between the input and output sequences to achieve *contextual alignment* between the input and output sequences.

14.2.0.1 The embedding layer

The input sequence is first tokenized into smaller units, such as words or subwords, referred to as *tokens*. These tokens are subsequently mapped to fixed-length vector representations, the token *embeddings*. Initially, embeddings are retrieved from a pretrained static lookup table and typically have dimensions of 512 or 768. However, unlike the static embeddings derived by older seq-to-seq models, the initial transformer embedding matrix M is subsequently *aligned* to the task in hand during model *training*.[5]

In contrast to RNN and LSTM-based models, which process input sequentially, the transformer can process the entire input sequence simultaneously. If executed

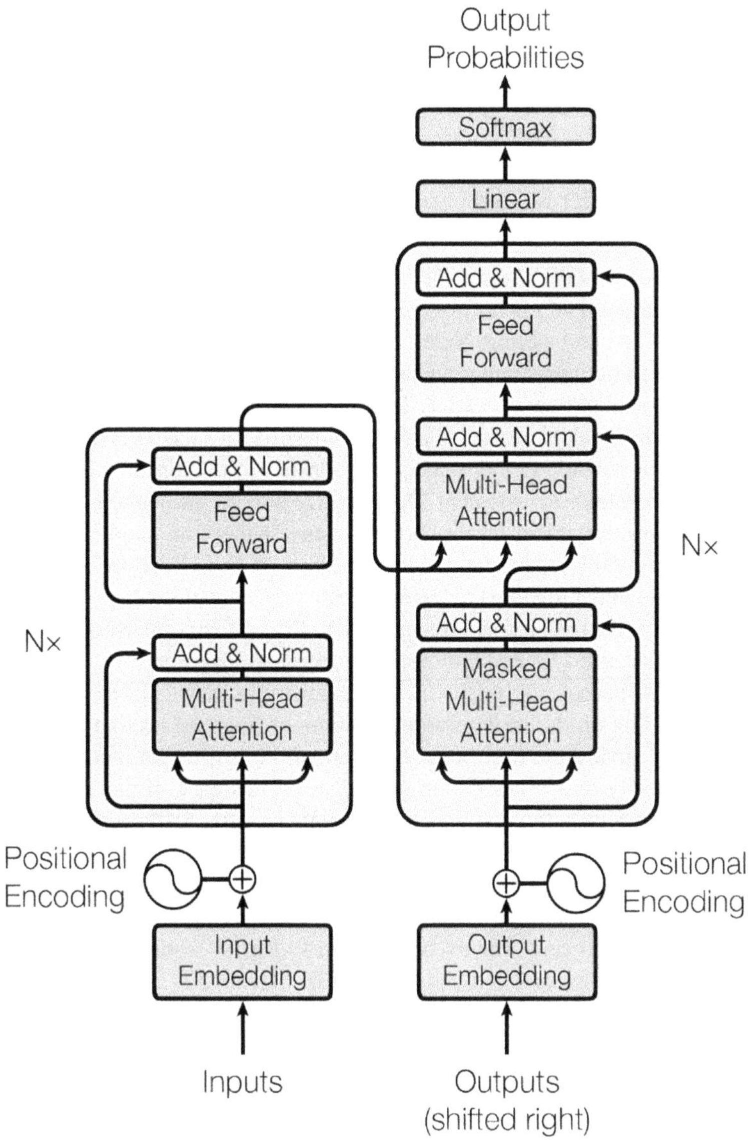

Figure 14.1 The transformer: model architecture; $N = 6$. (Image adapted from Vaswani, 2017 [358].)

with appropriate hardware in parallel, this significantly accelerates computation, thereby reducing training time. However, because transformers can process inputs in parallel, they lack the intrinsic word order information provided by sequential models, such as RNNs or LSTMs. To address this, positional encoding information is added to the token embedding to incorporate the position of each token within the sequence (see Section 14.1.2). The original proposal by Vaswani et al. utilized alternating sine and cosine functions to incorporate positional information into token embeddings.

14.2.0.2 The encoder blocks

A transformer encoder consists of the *embedding layer*, followed by multiple *encoder layers*. Each encoder layer comprises two sub-layers: a self-attention mechanism and a feed-forward neural network (FFNN). The input to the first encoder block, a sequence of token embedding vectors, is initially processed by the self-attention mechanism, which outputs an intermediate sequence of context-sensitive embedding vectors, as described in Section 14.1.3. These vectors are subsequently processed by an FFNN sub-layer, which consists of a fully connected neural network applied independently to each position in the sequence. It consists of two linear transformations with a ReLU activation in between. The first linear layer projects the input from the self-attention mechanism into a higher-dimensional space, enabling the model to capture more complex patterns and relationships. A ReLU activation introduces non-linearity, allowing the model to learn intricate representations. The second linear layer then projects the transformed output back to the original dimension, preserving the dimensionality of the embedding vector of the input tokens.

The number of neurons in the feed-forward layer of a transformer encoder block can vary, depending on the configuration. A common approach is to expand the dimensionality by having the first linear transformation project the input to a higher dimension, typically four times the original size. For instance, if the token embedding has 512 dimensions, the first linear layer might contain 2048 neurons (4×512).

Alternatively, the expansion factor can be treated as a hyperparameter, tuned based on the specific task and dataset. More significant expansions, such as eight times the input size, may increase expressive power but also lead to higher model size and computational cost. The exact number of neurons is often determined empirically through experimentation and tuning. The objective is to balance sufficient modelling capacity with practical considerations of model size and computational efficiency.

Thus, within the encoder block, it can be considered that the self-attention sublayer integrates information *horizontally* across the input sequence, allowing each token to *look at* others and capture dependencies and relationships, irrespective of their relative positions. In contrast, the feed-forward sub-layer operates on the embedding dimensions *vertically*, processing each token's embedding independently. This step refines the token's internal representation by combining and re-weighting features within its embedding vector.

Finally, *residual connections* and *layer normalization* (LN), while conceptually optional, are essential for numerical stability and convergence in transformer learning. The residual, or *skip* connections, add the input to the self-attention module to its output. Similar to the FFNN modules, layer normalization is applied independently to each embedding vector. The two standard methods are norm-last (post-LN) and norm-first (pre-LN) layer normalization. In norm-last, the output of each sub-layer is given by:

$$\text{LayerNorm}(x + \text{Sublayer}(x)),$$

... where $\text{Sublayer}(x)$ represents the function implemented by the sub-layer. In contrast, norm-first computes the output as:

$$x + \text{Sublayer}(\text{LayerNorm}(x)).$$

The original 2017 transformer employed the norm-last convention (see Figure 14.1). However, this approach proved challenging to train as it required careful hyperparameter tuning and a learning rate warm-up, where the rate starts small and gradually increases. In contrast, the norm-first convention, introduced in 2018 by Qiang Wang et al., [371], was found to be much easier to train, eliminating the need for warm-up and enabling faster convergence.

14.2.0.3 The decoder blocks

The transformer decoder process begins with an embedding layer. This layer takes the current output sequence, shifted to the right by one token, and mirrors the approach employed in the initial stage of the encoder, adding position information for each token in the input sequence. After the positional embedding, sequences are processed by a chain of N *decoder blocks* where the output of one decoder block forms the input to the next (see LHS of Figure 14.3; in Vaswani et al.'s original paper, [358] $N = 6$). Finally, a de-embedding layer is used to convert the output embeddings back to a sequence of tokens/words, ready to be passed back to the input, shifted right, and the process is iterated (see Figure 14.1).

Each decoder block comprises three sublayers: A masked multi-head self-attention(MMHSA) sub-layer, a non-masked multi-head *cross-attention* sub-layer, and an FFNN sublayer (see the RHS of Figure 14.3).

To prevent reverse information flow, the input sequence to the first sublayer of the decoder block is *partially masked* to ensure that it does not use future output to help it predict the next token. That is, the prediction for a token t depends only on the tokens from positions 1 to t (that is, the past and current tokens). A *causal mask* is constructed to achieve this behaviour before the Softmax stage by adding a triangular mask matrix M that is $-\infty$ at entries where the attention link must be cut and 0 at other places.[6]

Next, the decoder block also includes an additional attention mechanism known as *Cross-attention* or *encoder-decoder attention*. This mechanism allows the decoder to focus on the output embeddings generated by the encoder, all of which have been computed before the decoder begins its operation.

The intermediate representation produced by the MMHSA sub-layer serves as cross-attention queries (Q). At the same time, the encoder's output, which consists of contextually rich embeddings, provides both the keys (K) and the values (V). The cross-attention mechanism computes attention scores between the Queries from the

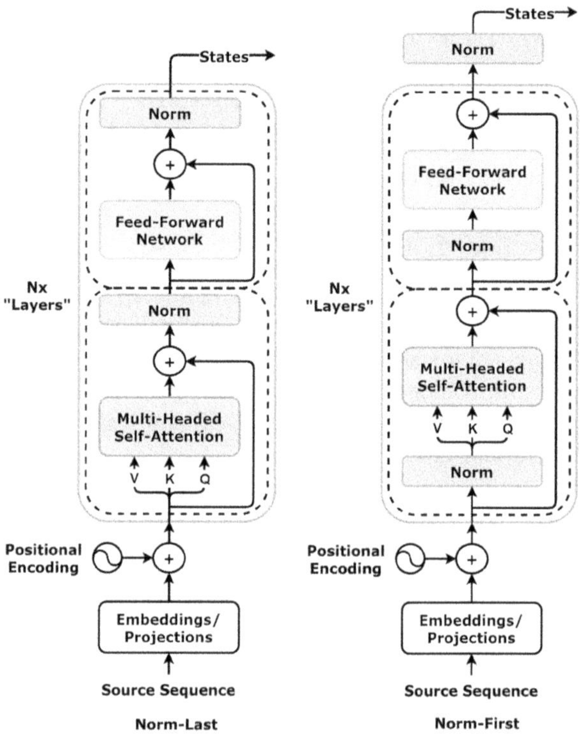

Figure 14.2 The transformer: encoder block showing norm-last and norm-first layer normalization. (Image adapted from Wikipedia: *transformer (deep learning architecture)* and licensed under the Creative Commons Attribution-Share Alike 3.0 Unported license: https://creativecommons.org/licenses/by-sa/3.0/deed.en.)

decoder and the Keys and Values from the encoder using the standard scaled dot product attention formula:

$$\text{Attention}(Q,K,V) = \text{Softmax}\left(\frac{QK^\top}{\sqrt{d_k}}\right)V,$$

...where d_k is the dimensionality of the keys.

These attention scores determine the degree to which each Query (from the decoder) focuses on each of the encoder's contextual embeddings. The result is a weighted combination of the encoder's values (V), tailored to the decoder's intermediate state. This mechanism enables the decoder to align and incorporate context from the input sequence while generating the output.

Since all contextual embeddings from the encoder are potentially relevant in generating the decoder output, no masking is required during the cross-attention process. Finally, after the cross-attention sub-layer and mirroring the pipeline in encoder blocks, an FFNN sub-layer is deployed to project the input from the cross-attention

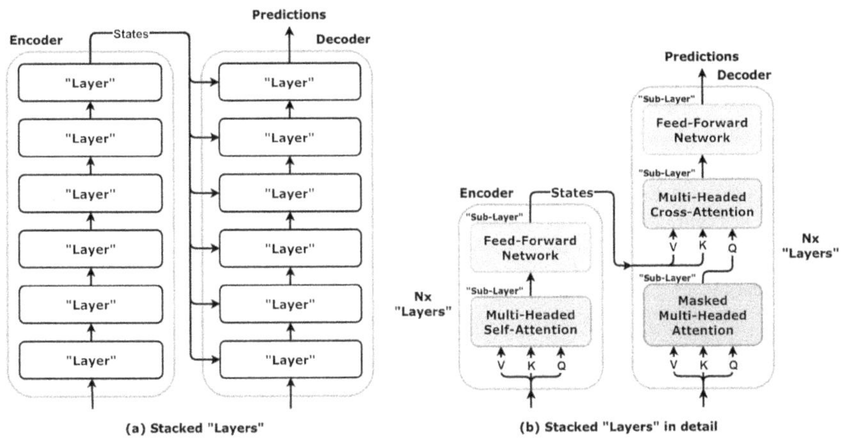

Figure 14.3 The transformer: decoder blocks. (Image adapted from Wikipedia: *transformer (deep learning architecture)* and licensed under the Creative Commons Attribution-Share Alike 3.0 Unported license: https://creativecommons.org/licenses/by-sa/3.0/deed.en.)

mechanism into a higher-dimensional space, enabling the model to capture more complex patterns and relationships and then project back to the original embedding dimension.

As per encoder blocks, decoder blocks also deploy residual/skip connections to bypass each of the three sub-layers (the multihead self-attention, cross-attention, and FFNN), each recombined using either pre- or post-layer normalization.

Following the final decoder block, a de-embedding layer generates output probabilities over the token vocabulary. A token is then sampled based on these probabilities, added to the decoder input shifted right, and the decoder is iterated to produce subsequent tokens seriatim, thereby generating the transformer output text autoregressively.

In summary, the decoding phase begins with the decoder receiving the contextually embedded input sequence and producing an initial output token. Subsequently, the decoder combines the encoded input with the newly generated output token to generate the next token. This iterative process continues until the decoder outputs a designated *stop* token. Finally, the sequence of tokens is mapped back into natural language, forming the transformer's output sequence.

The transformer architecture has several key advantages over older RNN-based models. First, since it does not need to process tokens sequentially, it can process relatively long input sequences more efficiently than RNNs.[7] Second, it is easier to parallelize the training process since the self- and cross-attention mechanisms allow the model to process different parts of a sequence independently. Finally, the transformer architecture has been shown to be very effective at capturing long-range dependencies in the input sequence, which is particularly important for natural language processing tasks.

Notes

1. We derive our material on transformers from Vaswani's seminal paper, [358] and two excellent, gentle introductions from Bradney Smith, *'LLMs from Scratch' series*, [328] and Arjun Sarkar 'All you need to know about attention and transformers'. [297]

2. In addition, related architectures for attention can be traced back to *fast-weight memories*, as explored by Jürgen Schmidhuber as early as 1992, [306] and the dynamics of *stochastic diffusion search*, as explored by Robert Summers in the context of human vision in 1998, [335] and J. Mark Bishop, *one of the authors of this book*, and colleagues in the SDP Research Group. [242] [4] [3]

3. Selecting values for d_q, d_k, and d_v smaller than d_{model} reduces dimensionality, potentially improving computational efficiency.

4. Jürgen Schmidhuber argues that, sensu stricto, the first transformer variant dates to 1991, with the publication of the *unnormalized linear transformer with linearized self-attention.* [303]

5. The embedding matrix M is a learnable parameter that maps discrete input tokens (words, subwords, or characters) into dense, continuous, real-valued vector representations by multiplying a one-hot vector representation of the token z—where z is vector of dimensionality v, i.e. the number of tokens in the vocabulary, with all entries being 0 except for a single 1 at the position corresponding to the specific token index labelling token z—with the embedding matrix M. This is a core component of the transformer, mapping symbolic linguistic data into a vector format that the model can process numerically.

   The embedding matrix M is a $V \times d$ matrix, where:

   - V: Number of tokens in the vocabulary.
   - d: Dimensionality of the embedding vectors (typically 512 or 768).

   Each row $\mathbf{m}_i$ of M represents the embedding vector for the i-th token in the vocabulary:

$$M = \begin{bmatrix} \mathbf{m}_1 \\ \mathbf{m}_2 \\ \vdots \\ \mathbf{m}_V \end{bmatrix}, \quad \mathbf{m}_i \in \mathbb{R}^d$$

   The one-hot vector $\mathbf{z}$ is a V-dimensional vector with all entries being 0 except for a single 1 at the index corresponding to the token. For example, if the token corresponds to the kth index in the vocabulary:

$$\mathbf{z} = \begin{bmatrix} 0 \\ 0 \\ \vdots \\ 1 \\ \vdots \\ 0 \end{bmatrix}, \quad \text{where the 1 is at position } k.$$

   When we multiply M $(V \times d)$ by $\mathbf{z}$ $(V \times 1)$, the result is a vector of dimensionality d $(1 \times d)$. The multiplication is defined as:

$$\mathbf{e} = M^\top \cdot \mathbf{z},$$

   ...where $\mathbf{e}$ is the resulting embedding vector of the token. This operation selects the k-th row of M (the embedding vector corresponding to the token at index k) because only the k-th entry of $\mathbf{z}$ is 1, and all other entries are 0. Thus:

$$\mathbf{e} = \mathbf{m}_k$$

   In this way, multiplying the embedding matrix M by the one-hot representation of token $\mathbf{z}$, retrieves the embedding vector for the token $\mathbf{z}$ directly.

6. For a sequence of length n, the mask is structured as follows:

$$\text{Mask} = \begin{bmatrix} 0 & -\infty & -\infty & -\infty & \dots & -\infty \\ 0 & 0 & -\infty & -\infty & \dots & -\infty \\ 0 & 0 & 0 & -\infty & \dots & -\infty \\ \vdots & \vdots & \vdots & \vdots & \ddots & -\infty \\ 0 & 0 & 0 & 0 & \dots & 0 \end{bmatrix}$$

$$\text{MaskedAttention}(Q,K,V) = \text{Softmax}\left(\frac{QK^{\top}}{\sqrt{d_k}} + M\right)V \tag{14.1}$$

$\dots$where $d_k = key\ size = dimension\ of\ key\ vectors$

After computing the attention scores, the product of dots between queries and keys, the casual mask is applied. For masked positions, the large negative value causes their corresponding entries in the attention matrix to become effectively zero after applying the Softmax function. This ensures that the decoder cannot attend to or derive information from future tokens.

7. Almost all of the high-profile advances in AI and deep learning have been made using models based on transformers due to their ability to learn at scale. However, the core building block of transformers, the attention operator, exhibits algorithmic quadratic cost as a function of sequence length, limiting the amount of context accessible. In 2023, Poli et al. proposed *Hyena*, a sub-quadratic drop-in replacement for attention (by interleaving implicitly parametrized long convolutions and data-controlled gating) that enables LLMs to be efficiently deployed, even on very large inputs. [273]

15 Large Language Models and Multimodal Systems

A large language model (LLM) is a machine learning model, typically based on a transformer architecture, that is constructed to generate natural language text from a prompt using large amounts of training data.[1] By analogy, imagine that an LLM is like a highly skilled librarian who has read millions of books in many languages. The librarian does not need to understand each word in every book because they have seen enough examples of how words and sentences usually fit together. When you ask a question or provide a prompt, the librarian does not simply look for an exact answer in one book. Instead, they use patterns and knowledge they have gained from all the books they have read to piece together the most relevant and coherent response. Similarly, LLMs use patterns from enormous amounts of data they've been pretrained on to generate cohesive and appropriate responses based on the prompts they receive. Although initially developed for natural language processing applications, the process of encoding words and tokens as embeddings can also be applied to other modalities (images, video, audio, etc.). This adaptability explains why LLMs are—as multimodal systems—particularly significant in the context of AVs.

15.1 THE GENERATIVE PRETRAINED TRANSFORMER

The GPT-3 175B model *davinci*, the largest version of GPT-3, employs a *decoder-only* architecture, i.e. it omits the encoder portion of the original transformer design. Compared to Vaswani's original transformer, the decoder block is simplified. It consists only of a multi-head self-attention mechanism, feed-forward neural network, residual connections and layer normalization. In contrast, Google's BERT, introduced by Jacob Devlin et al. in October 2018, adopts an *encoder-only* approach, disregarding the decoder component. [86]

'*Davinci*' deploys 96 attention heads, processing rich token embeddings of dimensionality 12,288, i.e. $d_{\text{embed}} = 12,288$. Each attention head enables the model to focus on different parts of the input sequence simultaneously, thereby capturing diverse relationships between tokens. The multi-head attention mechanism is deployed across 96 transformer decoder-block layers.

The *davinci* model was pretrained on vast, unlabelled data sets scraped from the internet, contributing to its ability to understand and generate sophisticated text responses to user prompts. In addition, it could be further fine-tuned for specific tasks such as sentiment analysis, text summarization, or question answering.

Both GPT and BERT, though inspired by the original transformer architecture introduced in the paper 'Attention is All You Need', [358] diverge from the complete encoder-decoder architecture because they are tailored for different tasks. The encoder-decoder framework was designed for specific sequence-to-sequence tasks,

DOI: 10.1201/9781003561460-20

such as machine translation, where the encoder processes the source sequence, e.g. English text, into a latent representation, and the decoder transforms this representation into the target sequence, e.g. French text. GPT and BERT do not require this dual-component transformer architecture, as their objectives differ.

GPT, typically used for language generation, relies solely on decoder-block layers. Its goal is to predict the next token in a sequence, given the preceding tokens, a process known as autoregressive modelling. The decoder's *causal masking* ensures that the model only attends to tokens up to the current point, enabling it to deliver a diverse range of functions such as text completion, creative writing, or storytelling. Since GPT does not involve transforming one sequence into another, an encoder is not necessary. This simplified, decoder-only approach reduces computational complexity and focuses the model's capability on text generation.

BERT, on the other hand, is optimized for *language understanding* tasks, such as question answering and sentiment analysis. It employs only the encoder, which uses bidirectional self-attention to process an entire sequence simultaneously, capturing context from both left and right. This capability enables BERT to comprehend nuanced relationships within the input text, making it well-suited for classification and comprehension tasks. Since BERT is not optimized for language generation, a latent sequence decoder is not required, further streamlining its architecture for its specific purpose.

The choice of architecture—encoder-only for BERT and decoder-only for GPT—reflects their distinct specializations. Encoder-decoder architectures, such as those in the original transformer or in models like *T5*, remain essential for tasks that require sequence transformations, including machine translation. By omitting unnecessary components, GPT and BERT achieve both computational efficiency and task-specific effectiveness, highlighting the adaptability of transformer-based designs.

15.2 LLMS FOR VISION

Andrej Karpathy, renowned for his work on Tesla's autopilot program, along with Li Fei-Fei, known for her contributions to ImageNet, adapted language processing techniques to image captioning, the task of generating textual descriptions of images. [176] This methodology subsequently influenced computer vision, and today, the use of LLMs underpins Tesla's *Autopilot* and *Full Self-Driving* (FSD) features.[2] Tesla employs generative pretrained transformers for vision on its TRIP chips, which allows the system to process sequential frames from video rather than relying solely on still images.

To further its self-supervised learning strategy, Tesla is developing an inference-optimized supercomputer called Dojo, specifically designed for neural network training. The Dojo system, based on Tesla's proprietary D1 chip, delivers 22.6 teraflops of FP32 performance, featuring 50 billion transistors, 10 TB/s of on-chip bandwidth, and 4 TB/s of off-chip bandwidth.

With active learning techniques, Tesla's system can selectively curate the most valuable video clips from a fleet of approximately 750,000 camera-equipped, internet-connected vehicles. Presently, Tesla operates its AI systems on 14,000 GPUs within its data centre, with access to a 30-petabyte video cache, which is expected to expand to 200 petabytes. Around 4,000 GPUs are dedicated to auto-labelling, while the remaining 10,000 support model training. The Dojo architecture will substantially enhance this performance, with the goal of further advancing Tesla's autonomous driving capabilities.

15.3 VISION TRANSFORMERS

'Vision transformers' (ViTs) apply the *transformer model*, originally developed for natural language processing, to computer vision tasks; however, while Vaswani et al. describe basic transformers for NLP tasks as consisting of both an encoder and a decoder, [358] akin to the GPT series of LLMs, the ViTs usually deploy an *encoder-only* architecture. Unlike CNNs, ViTs process images as sequences of patches using self-attention mechanisms; hence, the output of the encoder is passed to a neural network head to make an image *patch* prediction.

Following Dosovitskiy et al.'s original description, the input image is divided into fixed-size patches, e.g. 16x16 pixels, which are flattened and treated as tokens. [94] They describe reshaping an image of height H, width W, and channels C into N tokens with patch size P, where $N = HW/P^2$.

Tokens extracted from an image are often transformed using a learnable linear projection layer to adjust their length. This transformed length is known as the *latent dimension, channel dimension, or token length.* A learnable prediction, or classification, token—initially null—is prepended to the sequence. This token aggregates information from all patches during processing and is used to produce the final classification output. Since the core self-attention mechanism used in transformers is permutation-invariant,[3] positional encodings are added to these patches to preserve spatial information.

The position-embedded sequence, including the prediction token, is passed through several layers of the transformer. Each layer comprises a multi-head attention mechanism, which allows the model to compute relationships between patches and a feedforward network that processes the outputs of the attention mechanism. As usual, layer normalization and residual connections are applied to stabilize training and improve performance. In this way, the transformer's self-attention mechanism computes relationships between all patches, enabling the model to capture both local and global structures and features.

After passing through the transformer layers, the final representation of the prediction token is extracted and fed into a classification head, typically implemented as a linear layer, to predict the output class. The entire model is trained end-to-end using standard optimization techniques, such as backpropagation.

ViTs mark a significant departure from the localized feature extraction approach of CNNs, as they excel at capturing long-range dependencies and complex relationships between image regions more effectively than convolutional architectures. This capability makes ViTs highly effective for image analysis, classification, object detection, and segmentation, particularly when applied to large-scale data sets.

Hence, ViTs, although often requiring large data sets and significant computational resources for training, have shown promising applications in tasks relevant to autonomous driving. Gong et al. demonstrated this in their work on FUTR, where ViTs were applied to predict future actions from a video stream. [117] Similarly, FUTR3D, introduced by Chen et al., extended the use of ViTs to multimodal sensor fusion, enabling the integration of data from multiple sensors to enhance perception capabilities, further highlighting their relevance to autonomous driving. [58]

15.4 DIFFUSION MODELS

Diffusion models (DMs) are a class of generative AI models that create data samples by progressively refining random noise guided by a learned probabilistic process. They are at the heart of generative AI models such as *GLIDE, DALL-E by OpenAI, Imagen by Google, and Stable Diffusion*. Due to their complexity, only a relatively high-level description of their operation will be provided.[4]

The core principle underpinning DMs, inspired by non-equilibrium statistical physics, is to systematically and slowly destroy structure in a data distribution through an iterative forward diffusion process. Then, subsequently, to learn a reverse diffusion process that restores structure in data, yielding a highly flexible and tractable generative model of the data. A forward diffusion process iteratively corrupts data by adding Gaussian noise, and a reverse diffusion process reconstructs the data from noise. DMs are latent variable models that refer to a hidden continuous feature space, similar to VAEs (cf. Section 11.5.2), grounded in non-equilibrium thermodynamics.

By way of analogy, imagine sculpting a detailed statue from a block of clay. Diffusion models work like a sculptor, but in reverse. In the forward process, the sculptor starts with a finished statue and, thoughtfully and gradually, adds layers of clay until it becomes an unrecognisable blob. This process represents the *forward diffusion*, where data is incrementally corrupted until the statue becomes a random, unshapen, messy blob of clay.

The reverse process involves the sculptor working backwards, thoughtfully and gradually removing clay layer by layer until the original details of the statue are restored. At each step of the reverse process, the sculptor relies on her learned knowledge about what the statue *should* look like, ensuring the gradual refinements are accurate and consistent with her practice of sculpture.

Thus, described in computational terms, the forward diffusion process incrementally adds Gaussian noise to data over several steps, eventually transforming the input into pure noise. This process is parameterized to ensure each step can be inverted. The reverse diffusion process, which is the model's *primary learning objective*, involves learning to denoise the corrupted data step-by-step, reversing the forward process. At each time step, the model predicts the denoised data conditioned on the noisy input and the time step, progressively guiding the reconstruction toward the original data distribution.

Thus, just as the sculptor progressively shapes the messy blob of clay into a recognisable statue, the diffusion model refines random noise into coherent, high-quality data. In this way, diffusion models have gained significant attention for their ability to generate high-quality images, videos, and other types of data.

A significant similarity between diffusion models and transformers lies in their shared use of sequence processing techniques. In diffusion models, the denoising steps are often modelled using architectures that can handle sequential data, such as U-Net-based convolutional networks or transformer-like architectures. When using transformers in diffusion models, the self-attention mechanism enables the model to capture long-range dependencies between elements in the data, which is especially useful for high-dimensional data like images or videos.

Another conceptual parallel is the way positional or temporal information is encoded. Just as transformers use positional encodings to inform the model of the order of input tokens, diffusion models use time-step embeddings to provide the model with information about the current stage in the diffusion process. This temporal context is crucial for learning how to denoise effectively at each step. In addition, to make the diffusion model conditional on external input—e.g. a user prompt such as a text embedding—the output of a transformer attention mechanism is concatenated.

Diffusion models are highly relevant to autonomous driving, particularly in the areas of trajectory prediction and planning. They excel in predicting and planning vehicle trajectories in complex scenarios, as demonstrated by the *diffusion planner*, which models multimodal driving behaviour and facilitates cooperative interactions between vehicles. [390] These models generate diverse and realistic future trajectories, crucial for anticipating the movements of other cars.

In safety-critical scenario generation, diffusion-based approaches outperform traditional generative models by producing realistic and practical traffic scenarios for testing autonomous driving systems. This ensures a thorough evaluation of autonomous vehicle technologies in high-risk conditions. Additionally, some diffusion-based architectures support the joint modelling of prediction and planning tasks, leading to more cohesive decision-making. [217]

The controllable nature of diffusion models allows for the generation of trajectories that adhere to specific constraints, making them particularly valuable for creating diverse simulation scenarios during testing. [372] Multiagent motion prediction also benefits from this approach, as methods such as *MotionDiffuser* employ diffusion to model the joint distributions of future trajectories for multiple agents, capturing a wide range of possible outcomes in a permutation-invariant manner. [165]

Diffusion models enhance safety and adaptability by learning trajectory score function gradients and using flexible guidance mechanisms, enabling safer and more adaptive planning behaviours. Techniques such as *Diffusion-ES* combine diffusion with sampling-based search to optimize trajectories for complex reward functions, allowing for flexible driving behaviours beyond simple lane-following tasks. [389]

Additionally, diffusion models integrate data from multiple sensors, such as lidar, cameras, and radar, to construct dynamic representations of a vehicle's environment. This is crucial for functions such as path planning and collision avoidance. By modelling complex, multimodal distributions and generating diverse, realistic scenarios, diffusion models represent a significant advancement in trajectory prediction, scenario generation, planning, and safety optimization, making them indispensable tools for autonomous driving systems. [290]

15.5 MULTIMODAL MACHINE LEARNING AND SCALING

AI models like OpenAI's DALL-E, Google's Gemini, and DeepMind's Gato have demonstrated remarkable multimodal capabilities, from generating images to playing video games. In particular, DeepMind's Gato, a transformer model trained on a diverse range of data modalities, including images, text, proprioception, and joint moments, sparked considerable debate about its potential contribution to achieving artificial general intelligence, as its multimodal training enabled Gato to perform a variety of tasks with substantially increased accuracy. [122]

In addition to multimodal approaches, *scaling* has been a central theme in the pursuit of AGI. For example, researchers like Nando de Freitas have long maintained that larger models with more parameters will lead to a breakthrough in achieving AGI. The emphasis on scaling can be traced back to the success of transformer models like GPT-3, which demonstrated significant performance improvements with massively increased parameters compared to their predecessors. Equally, Gato's larger variants also demonstrated superior performance, further supporting the scaling hypothesis based on what is now known as scaling laws. [175]

While models like Gato indicate progress towards human-level AI, it is clear to many that scaling alone is not the solution. For example, researchers like Yann LeCun emphasize the importance of learning from the world, similar to how human babies develop understanding and prediction abilities. [128] Similarly, critics of LLMs, such as Gary Marcus, argue that scaling alone is insufficient and that new ideas and architectures will be necessary to move towards AGI. [222]

The academic debate surrounding large AI models, such as Gato, continues. While scaling *may* be a necessary step towards AGI, many believe it is not sufficient. Even OpenAI's CEO, Sam Altman, acknowledges the limitations of scaling and suggests that future progress will require exploring new approaches:

> 'We need another breakthrough. We can still push on large language models quite a lot, and we will do that. We can take the hill that we're on and keep climbing it, and the peak of that is still pretty far away. But, within reason, I don't think that doing that will (get us to) AGI. If (for example) superintelligence can't discover novel physics, I don't think it's a superintelligence. And teaching it to clone the behaviour of humans and human text—I don't think that's going to get there. And so there's this question which has been debated in the field for a long time: what do we have to do in addition to a language model to make a system that can go discover new physics?' [313]

As the field of AI evolves, researchers are shifting their focus beyond simply scaling up models. By exploring new techniques and approaches, they aim to unlock even greater capabilities and push the boundaries of what AI can achieve. Scaling up model size alone was insufficient for tasks such as arithmetic, common sense, and symbolic reasoning. One approach to extracting more world knowledge from multimodal LLMs (MLLMs) is by in-context few-shot learning via prompting. This approach is called chain-of-thought (COT) reasoning. [373] It is also deemed relevant for future AV applications.[5]

Next, we will examine how these innovations in deep learning have advanced the development of autonomous vehicles and what subsequent monetization strategies were derived from them.

Notes

1. For a detailed description, including many images using the example of ChatGPT, see Wolfram (2023). [386] For a broader and more layman-friendly presentation, see Heimann and Pummill (2023). [136]
2. Cf. Andrej Karpathy, Keynote Speech at Tesla Day 2021: https://www.youtube.com/watch?v=fdtC1AxFNkk
3. That is, the order of input elements (tokens or patches) does not inherently affect the model's ability to process them.
4. For details, readers are referred to 'Denoising Diffusion Probabilistic Models'. [142] This paper was the first to outline a *practical* framework for training and sampling from diffusion models, making them effective for generative tasks such as image synthesis.
5. In 2023, a Waymo study using MLLMs with *chain-of-thought* reasoning found: 'These results demonstrate that chain-of-thought reasoning can meaningfully improve driving performance, particularly when its components are carefully selected and integrated.' [159]

16 End-to-end Neural Networks

16.1 MODULAR VS END-TO-END APPROACHES

'End-to-end' neural networks have demonstrated outstanding success in extracting task-specific features and constructing representations across multiple domains, including natural language processing and gameplay. These models have already achieved superhuman performance in Atari video games and attained grandmaster-level results in complex multiplayer games such as *StarCraft* and *Dota 2*.

Moreover, as it typically does not require significant high-level reasoning, autonomous driving has long been considered well suited to an end-to-end approach. Indeed, as discussed on page 13, as early as 1988, Dean Pomerleau introduced the use of ALVINN, a three-layer end-to-end neural network controller, for road following.

In end-to-end vehicle control, the objective is to develop a control policy that directly converts sensor data into control commands. [193] Consequently, it is conceptually far more straightforward than a traditional modular stack, such as the one Stanley successfully deployed in the controlled setting of the DARPA Urban Challenge.

Building on this concept, in 2016, Nvidia advanced the approach for autonomous vehicles. [40] Instead of segmenting control tasks into separate modules—such as perception, prediction, and decision-making—the Nvidia end-to-end model employed a unified neural network that directly learned correlations between input images (pixels) and output controls (steering and acceleration instructions). In doing so, predicting driving commands such as steering angle, throttle, and braking directly from sensory input concomitantly enhances system efficiency.

Fundamentally, and in contrast to the modular control paradigm, end-to-end control closely mirrors human driving behaviour, in which perception and action co-occur. This approach marks a shift from an engineered system composed of data-driven components interconnected by extensive lines of code to a purely data-driven methodology based on a single end-to-end neural network. It exemplifies the subsequent realization of Richard Sutton's 'bitter' lesson: that, in the long run, methods with superior computational capability outperform those reliant on human knowledge and expertise.

16.2 TESLA'S END-TO-END APPROACH

Tesla's approach to autonomous driving evolved substantially with the introduction of Full Self-Driving (FSD) version 12 and subsequent versions. It shifted from a rule-based system to an end-to-end neural network model, sometimes referred to as *behaviour cloning*.

To summarise: before FSD 12, Tesla's autonomous driving relied, like the rest of the industry, on a hybrid system combining manually crafted rules with machine learning components. This involved explicit programming for various driving scenarios, supported by neural networks for tasks like object detection and lane recognition. Although effective to some degree, this approach had limited capacity to handle the complex variability of real-world driving conditions.

Obviously, end-to-end control requires a massive amount of data and computing power. It is easy to understand why Tesla embraces this concept—the company possesses a unique dataset of real-world driving data. The approach allows for a deep learning-only setup, which Tesla has advocated for some time. However, the most intriguing advantage for Tesla is that *self-supervised learning* promises to scale with data and available computing rather than human labour.

Tesla's potential dataset comprises approximately 20 million hours of 360-degree video per month across the fleet, and around 170 million hours across the eight cameras on each vehicle. Extracting this amount from the fleet and labelling the data manually would not be economically feasible for such a large amount of video footage. However, with custom supercomputer power coupled with intelligent algorithms, Tesla can extract interesting non-redundant sequences and automatically label them via self-supervised learning.

With FSD 12, Tesla transitioned to an end-to-end neural network architecture, marking a significant shift in its approach. In this new paradigm, the entire driving process—from perception through to decision-making—is managed by a single, unified neural network trained on extensive video data. This approach enables the system to learn complex driving behaviours and adapt to diverse scenarios without relying on manually coded rules.

The differences are evident. In terms of system architecture, prior to FSD 12, Tesla used a combination of rule-based logic and neural networks for specific tasks. In FSD 12 and beyond, Tesla employs a comprehensive end-to-end neural network that integrates perception, planning, and control. In terms of learning and adaptability, pre-FSD 12 relied on predefined rules with limited adaptability, requiring manual updates for new scenarios. Following Post-FSD 12, the system has further enhanced its adaptability through continuous learning from vast datasets, enabling it to handle a broader range of driving situations.

Tesla's pre-FSD 12 approach was constrained in performance and efficiency by the complexity of integrating rule-based systems with neural networks. According to a company statement, with the vehicle update to FSD 12, they removed the handcrafted software that formed the bulk of the vehicle's control module, eliminating more than 300,000 lines of C++ control code. This streamlining of the system concomitantly improves efficiency.

16.3 IMITATION LEARNING AND REINFORCEMENT LEARNING

Typically, two different learning methodologies are used in end-to-end approaches: imitation learning (IL) and reinforcement learning (RL). Both require human driving commands. Imitation learning is used in a supervised fashion and aims to learn a policy by observing the actions performed by human experts in example demonstrations. This approach has the advantage of being easily applicable to deep learning architectures. It optimizes the parameters of the model to reduce the difference between the model and human behaviour.

Reinforcement learning methodologies aim to learn a policy that maximizes the cumulative rewards received by the system as it interacts with the environment through exploration; whilst imitation learning can suffer from insufficient exposure to diverse driving situations, reinforcement learning is more resilient to this issue.

16.4 ADVANTAGES OF END-TO-END LEARNING

'End-to-end' approaches offer notable advantages in autonomous driving, including simplicity, efficiency, consistency, adaptability, and flexibility. This approach potentially enhances the robustness of the entire system.[1] Furthermore, and in contrast to traditional modular systems where predefined inputs and outputs of individual subsystems may limit adaptability across various driving scenarios, end-to-end models can dynamically attend to diverse information based on different road and traffic conditions.

Traditional modular stacks, which compress dynamic objects into 3D bounding boxes, restrict information retrieval for subsequent modules. Perception modules trained to detect these bounding boxes treat all objects equally, aiming to maximize average precision, even though nearby moving objects are most relevant to the driving task.

In contrast, end-to-end models trained through imitation learning can learn optimal intermediate representations directly from expert driving data. Without human-defined information bottlenecks, these models can leverage any implicit data in expert demonstrations, for example, detecting emergency lights reflected off surfaces, which may signal an obscured fire truck—a dangerous scenario that traditional systems, as deployed in, say, a Cruise AV, have often failed to interpret correctly.

Imitation learning also recognizes that human driving relies on contextual awareness beyond mere traffic rule adherence. For example, humans often slow down when they notice a child or an unleashed dog nearby. With *enough* training data, end-to-end models can similarly learn to account for complex, context-dependent scene factors that influence driving decisions, but in the real world, what is '*enough*'?

16.5 CHALLENGES TO END-TO-END LEARNING

Scaling an end-to-end control system using imitation learning presents challenges, primarily due to the difficulty of covering *all* potential scenarios within the training phase. Other challenges include extracting 3D scene structure from camera data, as this process relies on geometric cues, which can cause difficulties in differentiating between *moving* and *static* objects in certain scenes. In addition, factors such as rolling shutter effects, motion blur, and other transient objects further complicate the task in typical driving sequences.

However, a significant, broader concern with end-to-end learning is *interpretability*. Without intermediate outputs, tracing the root cause of an error or understanding the reasoning behind specific driving decisions is considerably more difficult, if not impossible. Modular systems, by contrast, allow engineering teams to focus on well-defined subtasks and independently refine the entire pipeline. This structure ensures system operability as long as intermediate outputs remain functional, with driving decisions emerging from a combination of subsystems operating on deterministic rules. These explicit intermediate representations make modular systems behave more predictably within their defined capabilities, facilitating more reliable reasoning about the decision-making process.

End-to-end models, on the other hand, function as black boxes.[2] Indeed, Wayve acknowledges explicitly that they lack visibility into the inner workings of their model. Silvius Rus, the company's software director, commenting on what drives the model's responses to new input, observed: 'It's like you're doing a brain scan, and you can see there's some activity in a certain part of the brain.' [135] And, while recent methods, such as attention maps, aim to improve interpretability by highlighting areas of algorithmic focus within a scene, the underlying issue remains, complicating safety and reliability assessments.

Moreover, open questions remain concerning design and architecture. Some approaches rely on deep learning architectures, assuming that all relevant driving knowledge can be learned directly from data. Others seek to design systems that are more similar to traditional, rule-based systems.

An additional design consideration is the choice of training method. Imitation learning, although widely used, is vulnerable to a challenge known as *distributional shift*.[3] This shift occurs when the state distributions of training and testing differ, limiting the agent's generalization. While reinforcement learning is an alternative, it requires extensive data, often millions or even billions of examples, to achieve convergence. As Yann LeCun wryly observed:

> 'A self-driving system for cars may require thousands of trials of reinforcement learning to learn that driving too fast in a turn will result in a bad outcome and to learn to slow down to avoid skidding. By contrast, humans can draw on their intimate knowledge of intuitive physics to predict such outcomes and largely avoid fatal courses of action when learning a new skill.' [198]

Notes

1. Pranav Singh Chiba and Pravendra Singha suggest end-to-end control could improve resilience against adversarial attacks, remarking: 'end-to-end driving demonstrates resilience to adversarial attacks. Adversarial attacks involve manipulating sensor inputs to deceive or confuse autonomous driving systems. In end-to-end models, it is challenging to identify and manipulate the specific driving behaviour triggers

as it is unknown what causes specific driving patterns.' [59] Nonetheless, successful adversarial attacks against end-to-end networks have already been demonstrated. [387]

2. There are some approaches on how to address this, for example, merging a text LLM with the driving algorithm. This produces insights but shifts the problem to the black box text LLM. Industry veteran Brad Templeton describes Wayve's approach as this: 'UK developer Wayve has merged an actual text LLM with their E2E driving system. You can ask it at any time why it's doing what it's doing. They hope that will make people feel better, as well as help debug it. When it was stopped at a red light with some cars ahead of it, I asked it why it wasn't going. It mentioned the red light but not the more important forward traffic, which I felt was a serious error since it wouldn't drive into them just because the light went green. This approach thus needs more work but may help deal with the fear.' [338]

3. De Haan, Jayaraman, and Levine identify another related issue, *causal confusion*. They provide the following example: in scenario A, a model's input includes an image of the dashboard and windshield; in scenario B, the same image is used but with the dashboard masked out. Although both models achieve low training loss, model B drives well in testing, while model A fails. This is because model A incorrectly learns to brake only when it sees the dashboard's brake light, mistaking the effect for the cause. [79]

Section VI

Two Routes to Autonomy

In the DARPA Challenges, two strategic approaches to autonomous vehicles emerged: the mapping-based strategy (the 'best map of the world') and the real-time computer vision approach (the 'gigantic eye'). Subsequently, Waymo pursued a *revolutionary* business model with traditional technology based on detailed maps and sensor fusion. In contrast, a relative newcomer, Tesla, has adopted a conventional business model with *revolutionary* technology; taking the 'gigantic eye' concept to the extreme and employing what we have termed the 'two eyes and a brain' strategy. This approach relies solely on simple cameras, with AI mimicking human perception, positing that artificial general intelligence is not just *necessary*, but is *sufficient* for autonomous mobility.

These technological strategies align with divergent business models. Google/Waymo has pursued a *revolutionary* path, seeking an immediate transition to driverless cars within a platform-based ride-hailing service. Tesla, by contrast, has adopted an *evolutionary* approach, incrementally enhancing driver assistance systems in privately owned vehicles, each with the eventual prospect of offering *robotaxi* capabilities. This creates a paradox: Tesla employs a conventional business model whilst advancing a revolutionary technological vision through AGI, whereas Google/Waymo seeks to disrupt the business model whilst relying on a traditional, modular technological framework derived from DARPA's approach.

Currently, both strategies continue to have their pro's and con's ...

17 A Question of Strategy

In the DARPA Challenges, two core concepts emerged: A mapping-based strategy (which we term 'the best map of the world') and a real-time computer-vision approach (which Sean Gerrish called the 'gigantic eye'). In the DARPA Urban Challenge, both methods were combined, as all teams used maps and partially supplemented them with real-time perception. Since all the DARPA approaches crucially relied on maps, we categorized them under 'best map of the world' in terms of their underpinning strategy.[1]

Another approach, especially popularized by Tesla in recent years, could be described as *computer vision and AI only*. Tesla took this to the extreme almost a decade after the Urban Challenge. We want to call this anthropomorphic approach the 'two eyes and a brain' strategy, referencing Elon Musk, who remarks that this is 'all humans need to drive'.[2]

17.1 THE BEST MAP OF THE WORLD

The fundamental principle of the *best map* approach is to create maximum redundancy through the interplay of different technologies. After Brooks, the basic theme of the *AI only* approach is to 'take the world as its best representation' and to perform all perception tasks in real-time with the most lightweight sensors available, e.g. mono cameras. In this regard, the approach shares similarities with human perception. Humans also do not use lidars, ultrasonics, and radars or detailed maps, but simply their eyes (which, as a pair, function as a stereo camera) and a brain (superficially, at least, akin to a neural network) combined with an intelligence that is able to generalize, reason, and to continually learn.

The DARPA Challenges had shown that AI was a *necessary* component of AVs. Tesla, on the other hand, was on its way to showing that AI was a *sufficient* component of autonomous mobility. This is a fundamentally radical approach to pursuing autonomy, as it requires AGI to handle all possible conditions that may unfold.

Interestingly, the two technological strategies were pursued within different business model paradigms that can be categorized as evolutionary (ownership-based *driver-assistance* model) versus revolutionary (access-based *robotaxi* model).

Google/Waymo aimed for an immediate leap to driverless cars, leveraging a platform-based car-sharing model. Tesla chose a step-by-step development, gradually evolving privately owned vehicles to ever more capable driver assistant systems that—eventually—gain robotaxi capabilities. Only at this point, where the automated systems would be driving safer than humans, could they be integrated into a dual-mode network of privately owned but eventually shared robotaxis, available for short-term lease through a Tesla-provided mobility app whenever their owners did not need them.

DOI: 10.1201/9781003561460-23

This observation highlights a paradox: Tesla adopted a conventional business model, producing and selling cars while aiming to revolutionize technology by achieving autonomous driving through AGI. Meanwhile, Google/Waymo sought to disrupt the business model itself, introducing a platform-based access economy while relying on a traditional technological framework rooted in DARPA's modular sense–plan–act approach with sensor fusion. Both companies aimed to disrupt the status quo but from fundamentally opposite directions. Which approach will prevail remains uncertain.[3] For the traditional automotive industry, accurately predicting the trajectory of this potentially transformative challenge is critical. Missteps could result in catastrophic financial losses, as evidenced by the examples of GM (later joined by VW) with Argo AI.

Disruptive innovation is a term coined by Harvard professor Clayton Christensen, who had carefully examined how new technologies can sometimes abruptly push longstanding products and services out of the market. Christensen identified a pattern: Digital substitutes had long appeared as inferior alternatives that could not compete with the more complex and mature products that prevailed. Due to these apparent limitations, incumbent players rejected digital approaches.

However, smaller startups—more agile and aggressive, with nothing to lose—were betting on digital and driving forward challenger products. As these technological products are built on highly scalable digital platforms, exponential advancements in development eventually propelled them into a dominant competitive position. Such technological breakthroughs can abruptly and permanently end the supremacy of product lines or even entire industries. Consider John Zimmer's example, in which online streaming services rendered DVD ownership and rental, *obsolete*—or, as Christensen described it, drawing from evolutionary biology, *disrupt* them.

The critical feature of modern information technology is that it is built on silicon chips, which to date, have consistently managed to double their performance roughly every twenty-four months—an effect described by Moore's law and an *exponential* function.

A prominent example of a *disruptive* market displacement is the substitution of wired telephony by *voice over IP* (VOIP) protocols. Initially, the quality of IP-based calls was unreliable, so incumbent manufacturers rejected the technology. However, by exponentially increasing network bandwidths, quality not only caught up but also delivered numerous other benefits, such as clearer voice quality and complete portability, at a much lower cost. This is precisely the kind of disruption to the century-old automotive industry, brought about by exponentially improving computer technology, that the tech players envisioned; nearly all of these were new to the mobility market.

In 2011, the influential internet pioneer and investor Marc Andreessen outlined the zeitgeist in a famous essay entitled 'Why Software Is Eating the World'. [11] The investor described how cloud computing prices had dropped by a factor of ten due to the spread of the internet infrastructure and how market entry barriers for new companies had virtually disappeared as a result. He then listed the numerous soft asset industries that had already been disrupted by software, including bookselling (including printing), film production and distribution, music, gaming, photography, marketing, recruiting, financial services, and education.

While these were all service industries that could be easily digitised, the investor warned that hard asset industries would not be spared from the trend, citing the automotive industry and autonomous driving as prime examples:

> 'Software is also eating much of the value chain of industries that are widely viewed as primarily existing in the physical world. In today's cars, the software runs the engines, controls safety features, entertains passengers, guides drivers to destinations and connects each car to mobile, satellite, and GPS networks. The days when a car aficionado could repair his or her car are long past, due primarily to the high software content. The trend toward hybrid and electric vehicles will only accelerate the software shift to electric cars that are entirely computer-controlled. And the creation of software-powered driverless cars is already underway at Google and the major car companies.' [11]

In contrast to Tesla, Google was pushing ahead with its *business model* disruption approach; in the robotaxi market, a driver was redundant. The passenger did not even need a driver's licence or any driving ability. To highlight this, Google deliberately made its first driverless ride with a passenger legally defined as blind.

However, Google's downsides were that they always had to map the drivable area, the *operational domain*, in great detail beforehand and that they still needed safety drivers to oversee the driving process. This severely limited the scope of their operation. Google's technology was thus both highly complex and very expensive.

A key advantage of Tesla's *technology* disruption approach is that its vehicles can be used off-the-shelf anywhere in the world—and that it can be utilized by Tesla to gather highly diverse global data from real-world driving conditions, invaluable for training and perfecting the AI networks; a key disadvantage is that the driver has to monitor the Tesla car at all times—so he or she needs full driving ability and licences, which currently rules out the use of the vehicle as a robotaxi. On the other hand, compared to Google's, Tesla's technology was comparatively inexpensive.

Clearly, the ultimate dream for autonomous driving would be a combination of Tesla's operational independence and Google's driver independence. However, both companies faced significant challenges. As Davies points out, initially, it was relatively easy for a small and talented team to create a car that could drive autonomously. [77] However, turning that into a commercially viable product that could handle every road, every situation, every time would require many years of work, billions of dollars, and hundreds, if not thousands, of workers, as engineers must robustly solve a seemingly endless series of edge cases, i.e. all the strange and unusual things that happen on the road, if it is possible *at all*.

Perfecting a technology within tightly controlled, specific conditions with a small team of dedicated developers is one thing; achieving real success in autonomous driving is quite another. It requires tackling labour-intensive challenges such as managing exceptions in a global context, ensuring ongoing maintenance, and guaranteeing seamless safety in operation. These demands also raise a host of new questions, as Davies points out—questions that engineers had not initially anticipated. For example, how much additional time and money would be needed to complete the work? What would 'finished' even look like? How good is 'good enough'? And how could the safety of the vehicle be assured? Furthermore, what would regulators, insurers, and the public make of the concept? Addressing all of these issues involves navigating some exceedingly tedious and complex paths.

In their quest for self-driving cars, Google faced unique challenges that Tesla did not. One of the biggest questions was how to finance their development on the way to market readiness. And how should they bring it to market? Should Google manufacture cars, partner with existing automakers, or offer a solution that individual drivers could install themselves? Furthermore, as they would find to their cost, developing the technology itself presented even more significant hurdles.

To function correctly, Google's self-driving cars needed incredibly detailed maps, generated by lidar laser scanning, which depict the environment down to a few centimetres. This was an enormous undertaking, especially considering that the maps needed to be regularly updated to account for changes such as shifting lane lines and the establishment of construction zones. Driving habits, rules, and conventions also vary from one region to another, making it difficult for a robot designed to drive in one area to adapt to another.

As a result, Google's only chance for success was to limit the complexity of the environment as much as possible. This meant working within a narrow *operational domain*, which restricted what the technology had to achieve by limiting the challenges it would face. All such limitations are expressed in the respective *operational design domain*.

For example, Google's Project Chauffeur team could be selective in choosing their missions to avoid dynamic obstacles like pedestrians and red lights or only operate in summer conditions. Although this strategy had limitations, it was seen as the best option for achieving progress in the field of self-driving cars. Thus, several ways emerged to reduce such environmental complexity; one approach was to limit the technology to a single type of road.

While the highway network is extensive—164,000 miles in the US alone—it is not so large that Google could not map it. Highway travel was relatively simple: a uniform direction of driving, structurally delineated opposing lanes, similar speeds for all road users, a uniform road surface, no intersections, no traffic lights, and no pedestrians or cyclists. Hence, Sebastian Thrun initially wanted to sell the technology to car manufacturers, who could sell it as a luxury add-on to their vehicles—a strategy later pursued by Ghost Autonomy.[4] In the city, customers could drive themselves; on the highway, they could hand over command to the car and do something else. With such an immediate return on investment, Google could further refine its system, eventually gaining the ability to master more complex situations.

Urmson and Levandowski had already discussed this option with carmakers in the US and Germany, but only met with disinterest. Ironically, Mercedes-Benz pioneered this approach in 2024 in collaboration with Nvidia, with BMW and Audi hot on its heels. In response to the rejection at the time, Levandowski considered simply selling the system as an aftermarket kit [77] that would enable any vehicle to steer itself on the highway—an idea that Cruise Automation, a company led by another DARPA Veteran, Kyle Vogt, was later to pursue for some time. [216] Today, this approach is mainly followed by Californian startup Comma.ai led by George Hotz, who was allegedly asked by Elon Musk in 2015 to build a software that could replace Tesla's MobileEye-based ADAS system.[5] Unlike Levandowski, however, Urmson wanted a more deeply integrated system.

Yet there was another reason that spoke against this form of operational simplification, and it weighed particularly heavily for the Google engineers: The better the technology worked, the less vigilant human drivers would be, rendering them the core risk factor in the process. Indeed, Alex Davies reports Dave Ferguson as critically observing:

> 'We realized that this was going to be much much harder to do than we had thought. It just became clear that look, if we can't rely on the driver to pay attention, or to hand over control, then basically, we just have to solve the whole problem. We need to have this vehicle not need the driver.' [77]

Above all, Google would object to any plan that deviated too much from the original objectives. Sergey Brin had aimed for a *moonshot*, not for becoming the auto industry's next supplier. In the end, their approach was to aim for potential disruption to car makers, not a way to sell incremental add-on features.

The cars needed to be fully autonomous, pushing along an urban mobility revolution and solving a problem that humanity faces, rather than just providing additional convenience to car owners within the existing paradigm. Google wanted to replace the horse and not improve the comfort of the carriage. With this directive, the other option to restrict environmental complexity was to only operate the vehicle in a carefully selected and confined geo-fence, which drastically limited monetization options.

Was Tesla's way superior to this? Not so clearly. It had the benefit of a clear business case with immediate financial returns and incremental investments. The technology was widely usable because the driver also acted as a safety operator. Furthermore, Tesla could use the presence of a human operator to harness invaluable real-world edge case data by deploying their *Shadow Mode*.

On the other hand, Tesla was burdened by the development of its vehicles. A task that Google had tried but quickly abandoned due to its high complexity and cost. Furthermore, while Google/Waymo has built a real and expanding robotaxi business, Tesla's plans for entry into this market—and a significant part of its fairy tale market valuation—entirely hinge on an unproven assumption about the development of AI.

Tesla's approach is an *all-in* bet on one card: *AGI*. Currently, its systems are not even Level–3 capable, not to mention Level–4. Hence, Elon Musk himself said (*and we return to this in 'The Amber Book'*):

> 'The overwhelming focus is on solving full self-driving. That's essential. It's really the difference between Tesla being worth a lot of money or worth basically zero.' [327]

Musk repeated the sentiment in June 2023, at the *Viva Technology Conference* in Paris, stating:

> 'Really, the value of the company is primarily on the basis of autonomy. If you look at our total vehicle output, it's almost 2 million vehicles this year or something like that. But that's still only 2 per cent of total vehicle production. The potential for autonomy is that the value of autonomy is so high that even if you have a discount, a percentage probability of autonomy happening, that is so incredibly valuable.' [121]

Furthermore, it remains an open question whether it is possible to reach full driverless autonomy by continuously evolving a driver-assistance system. Chris Urmson made this point—which is still valid—in 2015:

> 'Conventional wisdom would say that we'll just take these driver-assistance systems and well kind of push them and incrementally improve them, and over time, they'll turn into self-driving cars. Well, I'm here to tell you that, like me saying that if I work really hard at jumping, one day I'll be able to fly.'[6]

In addition, the *moonshot* approach could mobilize more endorsements from analysts at the time and spark more fantasies by venture capitalists than a supplier solution. Indeed, Google's way had already caught the attention of industry analyst John Casesa, whose observation of the changing market and the potential impact of self-driving cars on the auto industry was spot on.

Casesa recognized that the younger generations' buying habits were different and that traditional car manufacturers would need to adapt to survive. When Levandowski visited Detroit to share Google's vision of autonomous cars, he was met with scepticism. However, Casesa saw the potential of self-driving cars and recognized that they would not only disrupt but also replace traditional vehicles. That is, Casesa believed that Google's Project Chauffeur was a genuine substitute for existing technology rather than just an enhancement. For this open-minded investor, the self-driving car was akin to the transition from horse-drawn carriages to automobiles, and it would change the automotive industry forever. And this is precisely what caused him to engage.

However, Musk's driver-assistance approach with 'two eyes and a brain' was much more viable financially. After years of investment, Tesla is now profitable, with FSD serving as an additional revenue stream and providing an invaluable wealth of real-world data. In contrast, Waymo has high costs, relatively little turnover and a comparably modest market valuation. The vision of a market-dominating breakthrough into the robotaxi world, sparked by Tesla's alleged advancements in AGI, sent financial valuations soaring and made Elon Musk the wealthiest person in the world. This surpassed Larry Page, who had recognized the potential earlier but was constrained by complex technology and a slow expansion. The irony in this is that Page has seen the potential first (recall, Musk's attempt to sell Tesla to Page), that he was the one insisting on a disruptive moonshot with robotaxis and that it was Google that invented the business model of amassing data at any cost and figuring out later what to do with it. Indeed, the first instantiation of this idea was Google scanning the world's libraries for Google Books, with no business model in mind, simply because it was possible. Only later did they discover that this was an invaluable data source for other services, such as Google Translate, and led to the development of foundation models.

Although some analysts anticipate that if its growth moves forward as predicted, Waymo could reach revenue of 50–75 million while at the same time incurring losses of between \$1.2–\$1.5 billion, tech expert Brian Wang estimates: '*Waymo has probably had cumulative losses of \$15–20 billion since they started.*' [367] This is not a position that is easy to recover from.

In stark contrast, Tesla generates nearly \$100 billion in revenue today, makes a gross profit of around \$20 billion—breaking even in 2020—and has a market capitalisation that already surpassed \$1 trillion in 2021. Furthermore, Tesla has also outlined a clear plan to increase its valuation to \$2 trillion, which, ceteris paribus, would make it the sixth most valuable company in the world.

Nonetheless, unlike Tesla, Waymo has already proven that it can operate *driverless*. For Musk, *FSD Unsupervised* remains but an oft-repeated *promise*. However, with his focus entirely on AI, Musk placed his bet on exponential technology in Christensen's sense. Indeed, Musk precisely positions Tesla as a software company along the lines of Andreessen's 'software eats the world' mantra. In so doing, Musk had fully turned a massive spotlight on AGI-based self-driving capabilities, which he calls *generalized autonomy*, and this is precisely what makes him central to our trilogy.

Tesla has pushed to transform AI from a *necessary* to a *sufficient* condition for autonomous driving by facilitating the breakthrough to AGI in the robotics world. The company aims to develop its form of AGI to enable unrestricted autonomous driving in all environments (Level–5). If Tesla succeeds, the result will be a disruption of the mobility market that eclipses all previous industry changes and would far surpass OpenAI's recent plans for virtual *agentic* AI. [73]

17.2 TESLA'S ROAD TO EXPONENTIAL DISRUPTION?

As a direct consequence of Musk's faith in AI scalability, along their road to Level–5 autonomy, Tesla is discarding almost all that was learnt from the DARPA Challenges. Instead of highly specialized narrow AI systems, internal maps, and the fusion of multiple sensors, Tesla's 'two eyes and a brain' approach relies on cameras and a *generalised AI* based on a single end-to-end neural network.

While Sebastian Thrun revolutionized self-driving by making it a software problem, overturning the prevailing hardware-led paradigm, Elon Musk is taking a radical and decisive leap further by transforming self-driving into an AGI problem, upending the industry's narrow AI, modular approach. To achieve this ambitious goal, Tesla aims to execute the entire driving pipeline through vision and AI, even designing custom silicon to ensure the necessary real-time performance.

To process the visual data—and to illustrate Musk's point—Tesla turned away from convolutional neural networks, the very neural technology that helped make autonomy in real-world urban environments possible in the first place. In May 2024, Musk stated on X, in a remarkable war of words with CNN inventor Yann Le Cun, that Tesla does not 'use CNNs much these days', moving to other technologies, such as vision transformers, to power its vision pipeline.[7]

In the context of autonomous driving, compared to convolutional networks, vision transformers (and potentially diffusion models) offer superior global context awareness, which is crucial for understanding the scene as a whole; robust handling of complex interactions and occlusions within an image—a task critical in dynamic and cluttered driving environments; improved performance in multi-task learning, and scalability with large data sets.

No other company in the industry approaches self-driving cars like Tesla. Tesla's transformation into a vertically integrated AI and data company (owning the whole technology stack, from chip via car to cloud) puts it in a unique position. None of the competitors has established a comparable repeating cycle of creating data, tuning machine learning models, and deploying them on vast numbers of vehicles to generate additional data—a perpetuum mobile of big data prowess that would make any platform provider envious. If Tesla succeeds, it would be an impressive confirmation of Marc Andreessen's thesis that 'software eats the world' and that the automotive industry has all the prerequisites to premiere such *disruption* in the physical world.

To highlight the differences in data acquisition between the two approaches, it is helpful to compare the classic AV methods with Tesla's *big loop* data pipeline in more detail. In the conventional development of autonomous driving functions, a small test fleet of vehicles packed with sensors and hard disks is deployed to collect training data and map the environment.

The collected data for the 'best map of the world' approach consists of high-fidelity, three-dimensional lidar point clouds fused with data from stereo cameras and wide-range radar. This is necessary to build and maintain a high-definition map that provides a centimetre-precise representation of the complex urban environment, spatially accurate down to the smallest detail.

The map depicts features such as curbs, lane stripes, and pedestrian crossings, including the precise location of lane boundaries, lane widths, and the interconnection of lanes. The hard disks, which take up the car's entire boot,[8] store around 4–6 terabytes of data per hour of driving. Hence, the data-collecting capacity of a fully equipped test car can be up to 46 terabytes per drive, an amount of data that is so large that it cannot be uploaded to the cloud wirelessly or even via cable.[9]

Due to the sheer amount of data generated in test drives, hard disks must be manually extracted from the car after every drive and transferred to a high-capacity upload station near a large internet exchange point via a specialized mobile data centre. A new functional area, *data logistics*, emerged to manage this costly operation. For example, until April 2024, when their service was discontinued, AWS offered Snowmobile, a mobile data centre to physically move large amounts of data to a cloud access point.[10]

Hence, mapping the entire world—or even a country or just a city—is neither technically feasible nor logistically or commercially viable. This is one reason why the data-driven 'best map of the world' approach does not enable Level–5 autonomy. To cover all the roads in the world, even a one-off mapping effort would be far too time-consuming and costly.

It should also be noted that capturing an environment once is not enough. Environmental changes, such as those caused by road works, must be taken into account, as must different weather conditions, etc. Statistically, an average of 6,500 miles of new roads are constructed each year in the US alone. [1] Hence, data gathered in this way can only be used in a very tightly defined operational design domain to facilitate Level–3 or Level–4 autonomy.

Of course, the results of this vast data collection are not just used to map the environment but also to provide training data for AI algorithms. However, the approach has key downsides: To avoid overfitting, training data must be collected *en masse* and be diverse (*appropriately* varied). However, small fleets of test vehicles or robotaxis operating in a confined ODD gather data in uniform and largely repetitive driving situations, which are unlikely to be very diverse or contain edge cases. In addition, no simulation can be used to generate such edge cases, as they, by definition, mark *unknown unknowns* in the world.

So, what is Tesla doing? Well, it has a very clever answer. As Tesla does not rely on a 'best map of the world'—at least at the time of writing—it does not require specially equipped test vehicles to gather extensive geospatial sensor data. Instead, Tesla can leverage its entire fleet, currently exceeding 4 million cars worldwide, for data collection.

Since 2016, all new Tesla cars have been equipped with hardware capable of autonomous driving, although the full functionality is only enabled if the feature is subsequently purchased. Tesla understood early on that the data collected by its vehicles would be more valuable than the sale of optional upgrades like Autopilot or FSD—analogous to data-driven companies like Google and Facebook, which provide their services for free to users in order to collect their data. To harness the data from its global fleet, Tesla developed its crowd-sourcing approach called *Shadow Mode*. Thus, highly trained test drivers are not needed to gather data; in principle, every Tesla driver does.

When *Shadow Mode* is activated, the driver assistant feature does not actually control the vehicle; instead, it *shadows* driver behaviour in the background. As soon as the AV software predictions deviate from actual driving decisions, the computer can trigger the recording of a data snapshot for later upload. An *anticipatory* idea that resonates with Robert Rosen's *Anticipatory Systems*. [284]

Rosen suggested that human cognition involves internal models that simulate both the self and the environment, enabling the prediction of future events and proactive decision-making. Unlike reactive systems, humans adjust their behaviour based on these predictions, creating a continuous feedback loop of perception, cognition, and action. A loop which allows humans to continuously adapt their internal models as they interact with the world, improving future predictions.[11]

Similarly, Tesla's *Shadow Mode* continuously simulates the environment and only turns its attention to events that deviate from expected values. For data collection, this approach is doubly ingenious because it not only generates data that is likely to be an outlier—a valuable edge case—but it also generates data about a driving context that, by definition, the Autopilot or FSD system would not have handled appropriately had it been fully activated.

A critical side-effect of this methodology is that instead of manually and cost-intensively transporting petabytes of data to a physical high-speed data centre, not all of which may be relevant, human drivers help identify only the relevant data points; this massively reduces the amount of data collected and transmitted while massively increasing the value of each transmitted data point. Each day, Tesla sends this data to the cloud through the owner's private Wi-Fi connection, which, in any event, must be paired with the car to receive Tesla's routine software updates—a cheeky approach to externalize the cost of data acquisition that no traditional car manufacturer has dared pursue.

Tesla engineers can use this data set to retrain the neural networks and address gaps in their system, e.g. when a traffic sign is not identified because, say, a tree obscures it. In addition, Tesla has implemented another clever mechanism to help deal with these problem cases—it can send a query to its entire fleet that causes other vehicles to return *similar* data. For example, if a car is not recognized correctly because it has an unusual bike rack, a complete data set of cars with bike racks can be sourced from other vehicles in the field.

With its approach, Tesla collects more diverse data than any other automotive manufacturer. Some analysts view this data set as representing Tesla's actual value—a critical factor in understanding why, at their peak, Tesla shares were worth more than those of any other car manufacturer in the world.

With a fleet of approximately two million vehicles already equipped with what Tesla calls Full Self-Driving (FSD) hardware, Tesla's FSD fleet is already driving more real-world miles in a day (around seventy million) than Waymo's fleet has driven in its entire existence (around 20 million).

With this evolutionary (Level–2) strategy, Tesla has two critical advantages compared to Level–4 robotaxi providers: (i) they can operate in all *human driveable* areas (no ODD restriction), and (ii) they drive billions of miles on the road per year (26,6 billion, more than Waymo does in simulation) in very disparate real-world environments around the globe.

It is interesting to observe how Tesla pursues its approach to technological disruption. Let's remind ourselves how Clayton Christensen conceptualised disruption: Market entrants take cheap technology to find a foothold in less-profitable segments neglected by incumbents, and subsequently displace traditional competitors as soon as a technological leap makes their cheaper technology equivalent or better. [65]

This is precisely what Tesla is doing: Tesla deploys relatively inexpensive sensory hardware (simple mono-vision cameras), technology overlooked by the incumbent players. However, this method, in combination with massive data sets and advanced machine learning, could lead to the limitless applicability of autonomous driving in the event of a technological breakthrough—a conceptually simple but technologically radical idea.

17.3 MULTIMODAL END-TO-END NETWORKS AS A BREAKTHROUGH TO GENERALISED AUTONOMY?

The global investment firm ARK Invest, which manages the world's largest ETF fund, has developed an open-source model to predict the future performance of Tesla shares. [179] [14] Their forecast is based almost entirely on the assumption that AI and Tesla's superior data set constitute the *disruptive* factor—in the sense described by Christensen—propelling Tesla towards a near-monopoly in the driverless car market, encompassing both privately owned and fleet-managed vehicles.

ARK was founded in 2014 by analyst Cathie Wood. In 2019, she gained prominence for being the only analyst to correctly predict that Tesla's valuation would increase 'twenty or thirty times' by 2023. Since then, Forbes Magazine has referred to her as a 'superstar investor'.

ARK's success is built on its strategic investments in financial markets and disruptive technologies. Regarding Tesla, the firm is confident in its disruptive approach and predicts the launch of a robotaxi service within two years (from 2024). They estimate that the probability of Tesla not launching such a service within five years is *de minimis*. Consequently, ARK expects that by 2029, nearly 90% of Tesla's valuation will be driven by its robotaxi business. ARK explain the logic behind this claim as follows:

> 'With the release of Full Self-Driving (FSD) v12, Tesla transitioned to a neural network that learns from video data and directly controls the vehicle, eliminating over 300,000 lines of manual code. This update has enabled human-like—and in some instances, zero-intervention driving—by leveraging Tesla's significant data advantage, which now exceeds 1.3 billion cumulative FSD miles. According to our research, Tesla is accumulating data at $\sim 110\times$ the rate of Waymo.' [179]

What ARK does not mention here—and what is much less known in the public space—is that Tesla is not alone in this, nor even the first.[12] As we mentioned before, end-to-end neural networks for driving were pioneered by Pomerleau in 1988 for AV steering control and more comprehensively by Nvidia in 2016; technology that was adopted by the startup Wayve shortly after.

Today, in addition to Wayve, almost all self-driving companies deploy multimodal LLMs (MLLMs) in their stack one way or another: Waabi, Nuro, Momenta, Zoox or Pony.AI are some additional examples. [338] Even Waymo recently published research on end-to-end multimodal models using Google's Gemini MLLM, reporting promising improvements but also highlighting challenges and limitations.[13] [159]

Albeit, all players use MLLMs in different flavours—some alongside radar or lidar, some with lightweight HD maps, etc.

So, while Tesla was not the first, it could be argued that it has the most consequential and lightweight implementation, and it was undoubtedly the first to deploy the technology on the road at scale. More importantly, only Tesla vertically controls all elements in the stack to enable AGI-based self-driving, also known as *generalised autonomy*. Let's take a brief look at the technology and the hopes it sparks.

Erez Degan, President of Wayve, describes the situation like this:

> 'This approach uses a single, highly expressive AI model to convert raw vehicle sensing inputs directly into driving actions, eliminating the need for human-defined concepts to interpret sensor data. Central to this approach is self-supervised learning. By framing the learning problem as predicting the next data token in the sequence, whether it's the next word in LLMs or the next driving command in Wayve's model, the system thrives on unsupervised learning from raw, unlabelled data. The more data fed into the model, the richer and more expressive the AI model becomes for its specific application. The ability to train on a vast array of driving recordings without human input constitutes the power and magic of self-supervised learning.' [75]

Ghost Autonomy described the advantages in the following way:

> '[Multimodal] LLMs present the opportunity to solve [driving] *top-down*, starting from a world model. What if we could reason about driving with a model that is broadly trained on the world's knowledge and optimize it for executing the driving task? What if such a model could reason about a scene holistically, going from perception to suggested driving outcomes in a single step?' [189]

The alleged ability to reason (*sic*) is the main difference from that which has gone before, as it promises to deliver AI problem-solving similar to humans. Large language models have encoded generalised *world* understanding, which could be harnessed in novel situations. Instead of merely looking for road markers, as per CNNs, it is hoped that large models could actually develop an understanding of driving in lanes. This would provide the much-needed *abstraction*, the lack of which arguably plagues autonomous driving the most (we will revisit this in detail in the Amber Book).

This is to say: Proponents of this approach assert that going beyond pure probabilistic correlations and pattern matching, end-to-end learning provides a way to solve the problem of the long-tail—*the fundamental challenge of autonomous driving since its inception*—through reasoning and understanding the world in a human-like manner.

The lack of such capabilities is the reason why Level–5 autonomy is currently technically unfeasible and why even attention-free driver-assistance, in any broader scenario beyond a highway segment, is not possible. Or, as Prannay Khosla, former Ghost Autonomy Engineer and now founder of Stealth AI, a startup devoted to applying machine learning to real-world problems, explains:

> 'Attention-free/Level–4 autonomy requires a system that can reason about any given scene, even those never seen before, and deliver a safe driving manoeuvre every single time. The only models capable of this type of zero-shot generalization today are LLMs, finally delivering on the life-changing promise of autonomous driving.' [181]

Similar to humans, LLMs can utilize a substantial amount of contextual information in a process that facilitates higher-order decision-making within the driving system. This is the *fantasy* that currently fuels the AI hype, which is presently revitalising the autonomous driving industry. In their bullish scenarios, ARK expects Tesla's market capitalisation to grow from $540 billion to $10.9 trillion in 2029. In this scenario, its revenue would amount to an incredible $1.345 trillion, with $394 billion coming from electric vehicle sales and $951 billion from autonomous ride-hailing.

In October 2024, Tesla demonstrated two robotaxi vehicles without steering wheels and paddles, which it claimed would be produced by 2026, 'although he [Musk] said it could be as late as 2027'. [133] As a result, its business model would transform from one-off vehicle sales to a recurring revenue base '*as every car becomes an AI-powered cash-flow generation machine*'.

To get access to more data, Tesla launched a 30-day free trial in North America, which will add—at least temporarily—a large number of new data contributors. A rumoured FSD rollout in China would provide even more data sources. In its 2024 earnings report, Tesla showed prototypes of a ride-hailing app. This service could begin with human drivers and then gradually transition to a full robotaxi service.

Waymo has taken a more cautious approach to fully autonomous driving, focusing on careful, incremental improvements of the DARPA approach within clearly defined driving areas. It no longer pursues a moonshot strategy. The company recognizes that there will always be some limitations to their automation, mainly because their method is not easily transferable across different geographic regions.

Tesla's former chief AI scientist, Andrej Karpathy, has long argued that Waymo's reliance on high-definition maps makes their approach fundamentally unscalable. In contrast, he argues that Tesla's approach, which relies on computer vision and AI to manage vehicles in a manner analogous to human drivers, is the epitome of scalability.

Thus, Tesla's 'two eyes and a brain' approach to autonomous driving is fundamentally based on AGI. Tesla believes it is the only way to enable the vehicle to operate anywhere a skilled human driver could drive, thereby achieving Level–5 autonomy.

During his keynote speech at the 'Conference on Computer Vision and Pattern Recognition' (CVPR) in 2021, Andrej Karpathy explained Tesla's AI approach in detail. [91] Rather than using predefined maps, Tesla's technology learns about its surroundings in real-time, using video data from eight surrounding cameras—admittedly more than just 'two eyes'. The AI technology determines which areas are passable, where the lanes are, and the status of the traffic lights, all in real-time.

Furthermore, since 2019, Tesla has been busy designing proprietary silicon, which they term the 'Full Self-Driving chip'. This approach demonstrates Elon Musk's confidence in solving the problem of fully autonomous driving in the near future.

Indeed, Tesla is valued as a technology company rather than a vehicle manufacturer, in no small part due to its AGI strategy, which is highly rewarded by the capital market. Traditional vehicle manufacturers pale in comparison to Tesla; while Ford and GM each have a market capitalisation of around $42 billion, Tesla stands at a staggering $1.24 trillion in January 2025.

Tesla is skillfully adhering to the principles of disruptive innovation by fully embracing the exponential scalability of digital computers. This concept, introduced by John von Neumann in the early 1950s, is intrinsically linked to the idea of AGI and later connected to the *technological singularity*; a hypothesis is based on the observation that technology improves exponentially.

Computer technologies, for example, have doubled their power roughly every two years since 1950 (as described by Moore's law). Von Neumann and other experts have predicted that this rapid improvement in technology will eventually result in machine performance surpassing human capabilities. Today, OpenAI extended the principle into LLM training by observing that three factors scaled together—compute capacity, data set size, and number of parameters—lead to a function that resembles exponential growth. [175]

Such observations famously triggered Nando de Freitas—then Google DeepMind, now Microsoft—to assert that scaling is all that is required to reach AGI:

> 'It's all about scale now! The Game is Over! It's about making these models bigger, safer, compute efficient, faster at sampling, smarter memory, more modalities, INNOVATIVE DATA ...' [78]

Certainly, *if* computation can replicate all aspects of human ingenuity by scaling or other means, *then* AGI will be inevitable. Once AI surpasses human performance in all tasks, including driving, Elon Musk will finally realize his vision for Tesla, and there will be no inherent obstacle to achieving Norman Bel Geddes' 1939 vision of a fully autonomous mass transportation system.

Musk fully trusts this computational approach to intelligence, where the so-called 'scaling school' asserts that by scaling up data and processing power, emergent phenomena will begin to occur. In scaling up, emergence marks the point at which quantity shifts into quality, and the system demonstrates new capabilities. In this view, it is confidently assumed that *consciousness*, qualia, is just another such emerging quality. Musk's complete *faith* in computationalism also explains why he *believes* in things like mind uploading and why he fears that a singularity is imminent and poses the threat of a future super-intelligent AI enslaving humankind.

Notes

1. In reality, all AVs use some map, either pre-built or created on the fly. The main difference is the level of detail and whether it is stored or created ad hoc. Even Tesla uses maps to some extent. Waymo's cars,

on the other hand, are now perfectly capable of driving in areas that are not fully mapped, e.g. road works, albeit Waymo's overall approach is less map-centric now. In addition, although they are still using a modular system, they are now also using, or at least closely evaluating, the *LLM technology* invented by their sister company, Google and the *imitation learning* of their sister company, DeepMind.

2. Tesla used this statement to justify its avoidance of bulky lidar systems and the fusion of multiple sensor types, instead relying solely on cameras (which was merely a side component of the gigantic eye that was centred around lidar). Musk said this on several occasions. For example: 'Obviously, humans drive without shooting lasers out of their eyes. Unless you're Superman. But like humans drive just with passive visual, humans drive with eyes and a neural net and a brain neural net. So, the digital equivalent of eyes and a brain are cameras and digital neural nets or AI. So that's the entire road system was designed for passive optical neural nets. That's how the whole road system was not designed, and what everyone's expecting; that's how we expect other cars to behave. So, therefore, that is very obviously the solution for Full Self Driving ... as opposed to the very specific, neighbourhood by neighbourhood solution, which is very difficult to maintain, which is what our competitors are doing.' [339]

3. We want to highlight how presuppositional the assumptions of ARK Investment are given the uncertainty that surrounds Tesla's potential launch of a robotaxi any time soon from the perspective of early 2025, not only because of the unproven technical approach that would require another breakthrough in AI but also due to the challenging business case and scaling issues tied to the current narrow AI application and modular lidar and mapping-based approach. Furthermore, it is unclear whether the robotaxi can truly function as a flexibly deployable owned vehicle or whether it will need to be painstakingly operated in a fleet car-sharing mode in the future, as dual-mode cars are currently not even eligible for approval.

4. 'Attention-free driving starts on the highway—drivers are typically most interested in reclaiming long stretches of driving, say their daily commute or leisure trips, but that benefit only comes with attention-free self-driving, often called "hands-off, eyes-off" autonomy. By initially focusing on the highway commute domain, Ghost can achieve attention-free driving first for the driving domain that has the most value to consumers and auto OEMs.' [114]

5. 'Then he met with Elon Musk, who asked him to make a vision solution for self-driving cars that would rival the MobileEye tech Tesla was already working with.' [343]

6. 'How a driverless car sees the road'. A TED talk by Chris Urmson, March 2015: https://tinyurl.com/58na9xn7

7. Although Musk has stated otherwise, some trenchantly insist that Tesla will still need to rely on convolutional neural networks for translation-invariant feature extraction. [69]

8. USA: Trunk.

9. For example, uploading 50 terabytes over a 1 Gbps network would take around five days.

10. The AWS Snowmobile is a large-scale data transfer service provided by Amazon Web Services (AWS) that enables organizations to move vast amounts of data—up to 100 petabytes per Snowmobile—into the AWS cloud at a maximum upload speed of 1Tbps. Unlike other AWS data migration tools, such as AWS Snowball, which are portable data transfer devices, Snowmobile is a secure, truck-sized, air-conditioned mobile data centre housed in a secure shipping container.

11. Perhaps tellingly, in the context of Tesla's prospects for engineering *generalised intelligence* and Level–5 autonomy, Rosen also argued that this *anticipatory* ability in living systems goes beyond mere computation and cannot be fully explained by algorithmic processes alone.

12. ARK is obviously well aware that it is not only Tesla that is pursuing an end-to-end learning approach for self-driving cars. [178]

13. 'We demonstrate that EMMA (End-to-end Multimodal Model for Autonomous driving) can function as a generalist model in the autonomous driving domain, which jointly generates the outputs for multiple driving-related tasks. In particular, EMMA matches or even surpasses the performance of individually

trained models when it is co-trained with motion planning, object detection, and road graph tasks. Finally, we show EMMA's capacity to reason and make decisions in complex, long-tail driving scenarios. Despite these promising results, EMMA is not without its limitations. In particular, it faces challenges for real-world deployment due to (1) limitations in 3D spatial reasoning due to its inability to fuse camera inputs with lidar or radar, (2) the need for realistic and computationally expensive sensor simulation to power its closed-loop evaluation, and (3) the increased computational requirements relative to conventional models.' [159]

18 Green Lights Ahead!

The DARPA Challenges revealed that handcrafted, rule-based algorithms quickly reach their limits when faced with the complexity of an open world. More flexible, adaptive systems based on machine learning were needed, as statistics and probability proved more effective than logic and rules. The challenges also demonstrated the superiority of narrow AI approaches over more general ones *at that time*.

In the years following the final DARPA challenge, rapid advancements in computing and access to unprecedented amounts of data—*potentially encompassing the entire internet in the case of large language models*—fueled breakthroughs in AI. While earlier narrow AI systems, such as convolutional neural networks optimized for vision, were highly effective, they were still plagued with shortcomings stemming from their narrow nature—lack of appropriate generalization, zero-shot learning, understanding of common-sense physics, etc. The rise of generative AI, which facilitates more general and multimodal machine learning, has inspired hope that such limitations may eventually be overcome.

In autonomous driving, the central challenge is handling edge cases, which involves dealing with novel situations and, in turn, necessitates generalization. With sufficient data, perhaps transformers might be employed to create robust, general machine learning models that one day may equip machines with a world model comparable to human common sense—a concept regarded by researchers like Yann LeCun as the critical enabler of *human-level AI*. This development raises the prospect that artificial general intelligence could soon become achievable by appropriately scaling up these recent technological advancements.

With human-level AI, autonomous driving in any environment where a human can operate a vehicle would certainly be possible. The first company to achieve AGI would profoundly disrupt society, particularly the mobility industry, and ARK Investment's optimistic forecasts may well be swiftly realized. In this context, we observed that Tesla's entire approach is heavily focused on generalised autonomy and the pursuit of human-level intelligence, leveraging the *unreasonable effectiveness of data*, statistical methods, and generative AI.

Tesla is designed for scale, with custom-built chips and a strong focus on data acquisition and AI advancement. Using a single end-to-end neural network and simple vision camera to control the entire driving process, Tesla avoids the need for multi-layered control architectures, diverse sensors requiring data fusion, costly HD maps or complex rule-based interactions.

In the *Amber* volume of our *Driving Intelligence* trilogy—**Errors in Motion, or How Autonomy Fails**—we aim to assess two distinct approaches to autonomy: the 'best map in the world' versus 'two eyes and a brain', tested *in the wild*. Indeed, as we established at the outset, our objective is precisely this: to employ autonomous

DOI: 10.1201/9781003561460-24

driving technologies in real-world conditions as a precise benchmark for evaluating progress towards *human-level* artificial intelligence; to highlight what remains lacking and to explore whether these missing elements can ever be appropriately realized *in silico*.

Bibliography

1. Federal Highway Administration. Our nation's highways: 2011, 2011. Accessed: 2025-01-08. URL: https://www.fhwa.dot.gov/policyinformation/pubs/hf/pl11028/.
2. Igor N. Aizenberg, Naum N. Aizenberg, and Joos P. L. Vandewalle. *Multi-Valued and Universal Binary Neurons: Theory, Learning and Applications*. Kluwer Academic Publishers, Boston, MA, 2000. URL: https://link.springer.com/book/10.1007/978-1-4757-3115-6, doi:10.1007/978-1-4757-3115-6.
3. Mohammad Majid al Rifaie and John Mark Bishop. Swarmic paintings and colour attention. In *Evolutionary and Biologically Inspired Music, Sound, Art and Design: Second International Conference, EvoMUSART 2013, Vienna, Austria, April 3-5, 2013. Proceedings*, volume 7834 of *Lecture Notes in Computer Science*, pages 97–108. Springer, 2013. URL: https://link.springer.com/chapter/10.1007/978-3-642-36955-1_9, doi:10.1007/978-3-642-36955-1_9.
4. Mohammad Majid al Rifaie and John Mark Bishop. Swarmic sketches and attention mechanism. In *Evolutionary and Biologically Inspired Music, Sound, Art and Design: Second International Conference, EvoMUSART 2013, Vienna, Austria, April 3-5, 2013. Proceedings*, volume 7834 of *Lecture Notes in Computer Science*, pages 85–96. Springer, 2013. URL: https://link.springer.com/chapter/10.1007/978-3-642-36955-1_8, doi:10.1007/978-3-642-36955-1_8.
5. Igor Aleksander. Some psychological properties of digital learning nets. *International Journal of Man-Machine Studies*, 2(2):189–212, 1970. URL: https://www.sciencedirect.com/science/article/pii/S0020737370800281, doi:10.1016/S0020-7373(70)80028-1.
6. Igor Aleksander and Helen Morton. *An Introduction to Neural Computing*. Chapman & Hall, 1995. URL: https://dl.acm.org/doi/10.5555/131391.
7. John Aloimonos, Isaac Weiss, and Amit Bandyopadhyay. Active vision. *International Journal of Computer Vision*, 1(4):333–356, 1988. URL: https://doi.org/10.1007/BF00133571.
8. Shun-Ichi Amari. Learning patterns and pattern sequences by self-organizing nets of threshold elements. *IEEE Transactions on Computers*, C-21(11):1197–1206, November 1972. URL: https://ieeexplore.ieee.org/document/1672070, doi:10.1109/TC.1972.5009075.
9. Anil Ananthaswamy. *Why Machines Learn: The Elegant Math Behind Modern AI*. Dutton, New York, 2024. URL: https://www.penguinrandomhouse.com/books/677608/why-machines-learn-by-anil-ananthaswamy/.
10. Chris Anderson. The end of theory: The data deluge makes the scientific method obsolete, June 2008. URL: https://www.wired.com/2008/06/pb-theory/.

11. Marc Andreessen. Why software is eating the world, August 2011. URL: https://a16z.com/why-software-is-eating-the-world/.
12. University of Bristol Anonymous. Grey Walter and his tortoises, 2008. Accessed: 2025-01-03. URL: https://www.bristol.ac.uk/news/2008/212017945378.html.
13. Aqeel Anwar. Difference between autoencoder (ae) and variational autoencoder (vae), November 2021. URL: https://medium.com/data-science/difference-between-autoencoder-ae-and-variational-autoencoder-vae-ed7be1c038f2.
14. ARK Invest. Ark invest: Tesla valuation model, 2024. URL: https://github.com/ARKInvest/ARK-Invest-Tesla-Valuation-Model.
15. ARK Invest. Tesla is poised to launch a multi-trillion dollar robotaxi opportunity in 2025, October 2024. 28th October 2024. URL: https://www.ark-invest.com/newsletters/issue-437.
16. Countess of Lovelace Augusta Ada King. Notes by the translator: Note a. In Luigi Federico Menabrea, editor, *Sketch of the Analytical Engine invented by Charles Babbage, Esq.* Richard and John E. Taylor, London, 1843. URL: https://www.gutenberg.org/ebooks/75107.
17. Francis Bacon. *Novum Organum, sive Indicia Vera de Interpretatione Naturae*. John Bill, London, 1620. Published in Latin as part II of The Instauratio Magna. URL: https://www.gutenberg.org/ebooks/45988.
18. Dzmitry Bahdanau, Kyunghyun Cho, and Yoshua Bengio. Neural machine translation by jointly learning to align and translate. In *3rd International Conference on Learning Representations (ICLR'15)*, May 2015. 7–9th May. URL: http://arxiv.org/abs/1409.0473.
19. K. D. Baker and G. D. Sullivan. The knowledge-based approach. In *Proc. BMVC 1987*, 1987. URL: https://bmva-archive.org.uk/bmvc/1987/avc-87-001.pdf.
20. Anton Bakhtin, Hengyuan Hu, Weston Gray, Derek Sculley, Aurko Roy, Adam Lerer, Yonatan Zhou, Edward Lockhart, Tianqi Chen, Douwe Kiela, Mike Lewis, Kilian Weinberger, and Jason Weston. Human-level play in the game of diplomacy by combining language models with strategic reasoning. *Science*, 378(6619):1067–1075, November 2022. URL: https://noambrown.github.io/papers/22-Science-Diplomacy-TR.pdf, doi:10.1126/science.ade9097.
21. D. H. Ballard. Animate vision. *Artificial Intelligence*, 48:57–86, 1991. URL: https://www.sciencedirect.com/science/article/abs/pii/0004370291900804#:~:text=Animate%20vision%20systems%20have%20gaze,the%20larger%20context%20of%20behavior.
22. Daniel Bartz. Autonomous cars will make us safer, November 2009. Accessed: 2025-01-08. URL: https://www.wired.com/2009/11/autonomous-cars/.
23. Pritam Basu. What is deep learning .... Medium, March 2018. 10th March 2018. URL: https://medium.com/@basu.preetam/what-is-deep-learning-62ea755f8c03.

24. Paul Beattie and J. Mark Bishop. Self localisation in the SENARIO autonomous wheelchair. *Journal of Intelligent & Robotic Systems*, 22:255–267, 1998. URL: https://link.springer.com/article/10.1023/A:1008033229660.
25. Yoshua Bengio, Pascal Lamblin, Dan Popovici, and Hugo Larochelle. Greedy layer-wise training of deep networks. In *Advances in Neural Information Processing Systems 19: Proceedings of the 2006 Conference*, pages 153–160. MIT Press, 2007. URL: https://proceedings.neurips.cc/paper/2006/file/5da713a690c067105aeb2fae32403405-Paper.pdf.
26. Max Bennett. *A Brief History of Intelligence: Evolution, AI, and the Five Breakthroughs That Made Our Brains*. HarperCollins, 2023. URL: https://harpercollins.co.uk/products/a-brief-history-of-intelligence-why-the-evolution-of-the-brain-holds-the-key-to-the-future-of-ai-max-bennett?variant=41066050846798.
27. Nirav Bhan. A new ai-inspired definition of 'cope', January 2025. URL: https://www.linkedin.com/posts/nirav-bhan-b0b75a52_i-recently-learned-the-term-cope-and-it-activity-7276967187683115008-RaDd/.
28. Celeste Biever. ChatGPT broke the turing test — the race is on for new ways to assess AI. *Nature*, 619(7971):686–689, July 2023. URL: https://www.nature.com/articles/d41586-023-02361-7, doi:10.1038/d41586-023-02361-7.
29. T. O. Binford. Visual perception by computer. In *IEEE Conference on Systems and Control*, Americana Hotel, Miami Beach, Florida, December 1971. URL: https://www.bibsonomy.org/bibtex/1d3ccfd0bd4a73c88baeacf702478560a/schaul.
30. J. M. Bishop. Stochastic searching networks. In *Proceedings of the First IEE International Conference on Artificial Neural Networks*, pages 329–331, London, UK, October 1989. URL: https://ieeexplore.ieee.org/document/51986.
31. J. M. Bishop. Is anyone home? A way to find out if AI has become self-aware. *Frontiers in Robotics and AI*, 5:17, 2018. URL: https://www.frontiersin.org/journals/robotics-and-ai/articles/10.3389/frobt.2018.00017/full, doi:10.3389/frobt.2018.00017.
32. J. Mark Bishop. *Anarchic Techniques for Pattern Classification*. PhD thesis, University of Reading, 1989. URL: https://dl.acm.org/doi/abs/10.5555/915801.
33. J. Mark Bishop. History and philosophy of neural networks. In Hisao Ishibuchi, editor, *Computational Intelligence*. Encyclopedia of Life Support Systems (EOLSS), UNESCO, 2015. URL: https://www.eolss.net/outlinecomponents/computational-intelligence.aspx.
34. J. Mark Bishop. What game should artificial intelligence take on next? *New Scientist*, March 2016. 16th March. URL: https://www.newscientist.com/article/2081021-what-game-should-artificial-intelligence-take-on-next/.
35. J. Mark Bishop. Artificial intelligence is stupid and causal reasoning will not fix it. *Frontiers in Psychology*, 11:513474, 2021. URL: https://www.frontiersin.org/journals/psychology/articles/10.3389/fpsyg.2020.513474/full, doi:10.3389/fpsyg.2020.513474.

36. J. Mark Bishop, Paul R. Minchinton, and Richard J. Mitchell. Real-time invariant grey level image processing using digital neural networks. In *Proceedings IMechE Eurodirect 1991*, pages 187–188, London, UK, 1991. URL: https://www.researchgate.net/profile/John-Bishop-4/publication/240207825_Real_Time_Invariant_Grey_Level_Image_Processing_using_Digital_Neural_Networks/links/54862dda0cf289302e2c063a/Real-Time-Invariant-Grey-Level-Image-Processing-using-Digital-Neural-Networks.pdf.
37. Mark Bishop and Yasemin J. Erden, editors. *The 6th AISB Symposium on Computing and Philosophy: The Scandal of Computation - What is Computation?*, University of Exeter, April 2013. URL: https://philpapers.org/archive/MLLWIA-2.pdf.
38. W. W. Bledsoe and I. Browning. Pattern recognition and reading by machine. In *Proceedings of the Eastern Joint Computer Conference*, pages 225–232, 1959. URL: https://dl.acm.org/doi/10.1145/1460299.1460326.
39. Jon Bohren, Tully Foote, Jim Keller, and Alex Kushleyev. Little Ben: The Ben Franklin racing team's entry in the 2007 DARPA urban challenge. In *The DARPA Urban Challenge*, pages 231–255. Springer, 2009. URL: https://onlinelibrary.wiley.com/doi/abs/10.1002/rob.20260, doi:10.1007/978-0-387-98167-5_14.
40. Mariusz Bojarski, Davide Del Testa, Daniel Dworakowski, Bernhard Firner, Beat Flepp, Prasoon Goyal, Lawrence D. Jackel, Mathew Monfort, Urs Muller, Jiakai Zhang, Xin Zhang, Jake Zhao, and Karol Zieba. End to end learning for self-driving cars. arXiv: [cs.CV], April 2016. 25th April 2016. URL: https://arxiv.org/abs/1604.07316, arXiv:arXiv:1604.07316.
41. Jonathan Booher, Khashayar Rohanimanesh, Junhong Xu, Vladislav Isenbaev, Ashwin Balakrishna, Ishan Gupta, Wei Liu, and Aleksandr Petiushko. CIMRL: Combining IMitation and Reinforcement Learning for safe autonomous driving. arXiv: [cs.LG], June 2024. 13th June 2024. URL: https://arxiv.org/abs/2406.08878, arXiv:arXiv:2406.08878.
42. J. Borenstein, S. Levine, and Y. Koren. The navchair - a new concept in intelligent wheelchair control for people with multiple handicaps. In *Fifth Annual Conference on Technology and Persons with Disabilities*, pages 117–124, Los Angeles, California, 1990.
43. Bernhard E. Boser, Isabelle M. Guyon, and Vladimir N. Vapnik. A training algorithm for optimal margin classifiers. In *Proceedings of the 5th Annual Workshop on Computational Learning Theory (COLT'92)*, pages 144–152. ACM, 1992. URL: https://dl.acm.org/doi/10.1145/130385.130401.
44. Ronald J. Brachman, David Gunning, and Murray Burke. Integrated AI systems. *AI Magazine*, 41(2):66–82, 2020. URL: https://ojs.aaai.org/aimagazine/index.php/aimagazine/article/view/5300, doi:10.1609/aimag.v41i2.5300.
45. Leo Breiman, Jerome Friedman, R.A. Olshen, and Charles J. Stone. *Classification and Regression Trees*. Wadsworth and Brooks/Cole Advanced Books & Software, Monterey, CA, 1984. URL: https://www.taylorfrancis.com/books/

mono/10.1201/9781315139470/classification-regression-trees-leo-breiman-jerome-friedman-olshen-charles-stone.
46. Selmer Bringsjord, Paul Bello, and David Ferrucci. Creativity, the turing test, and the (better) lovelace test. *Minds and Machines*, 11(1):3–27, 2001. URL: https://link.springer.com/article/10.1023/A:1011206622741, doi:10.1023/A:1011206622741.
47. Terence Broad. Autoencoding video frames. Master's thesis, Goldsmiths, University of London, UK, 2016. URL: https://www.academia.edu/25585807/Autoencoding_Video_Frames.
48. Rodney A. Brooks. Intelligence without representation. *Artificial Intelligence*, 47(1):139–159, 1991. URL: https://www.sciencedirect.com/science/article/abs/pii/000437029190053M.
49. D. S. Broomhead and D. Lowe. Multivariable functional interpolation and adaptive networks. *Complex Systems*, 2(3):321–355, 1988. URL: https://sci2s.ugr.es/keel/pdf/algorithm/articulo/1988-Broomhead-CS.pdf.
50. Meredith Broussard. *Artificial Unintelligence: How Computers Misunderstand the World*. The MIT Press, kindle edition edition, 2018. URL: https://mitpress.mit.edu/9780262537018/artificial-unintelligence/.
51. Herbert Bruderer. Leonardo's self-driving car, July 2019. Accessed: 2025-01-08. URL: https://cacm.acm.org/blogcacm/leonardos-self-driving-car/.
52. William H. Calvin. *The Cerebral Code: Thinking a Thought in the Mosaics of the Mind*. MIT Press, Cambridge, MA, 1996. URL: https://mitpress.mit.edu/9780262531542/the-cerebral-code/.
53. Murray Campbell, A. Joseph Hoane, and Feng-hsiung Hsu. Knowledge discovery in deep blue. *Communications of the ACM*, 42(11):65–72, 1999. URL: https://dl.acm.org/doi/10.1145/319382.319396, doi:10.1145/319382.319398.
54. Katie Canales. Ousted uber co-founder Travis Kalanick has reportedly spent $130m on his'ghost kitchen' startup. here's what it's like inside one of the secretive locations, 2019. URL: https://www.businessinsider.com/cloud-kitchens-travis-kalanick-san-francisco-location-address-pictures-2019-11.
55. Giuseppe Pio Cannata. Vanishing gradient in deep neural networks: Causes and possible solutions. *Towards Data Science*, September 2021. URL: https://towardsdatascience.com/vanishing-gradient-in-deep-neural-network-83953217c59f.
56. Ernst Cassirer. *An Essay on Man: An Introduction to a Philosophy of Human Culture*. Yale University Press, New Haven, CT, 1944. URL: https://www.jstor.org/stable/j.ctv1mgmd44.
57. Zhanxiang Chai, Tianxin Nie, and Jan Becker. *Autonomous Driving Changes the Future*. Springer Nature Singapore, 2021. URL: https://link.springer.com/book/10.1007/978-981-15-6728-5.
58. Xuanyao Chen, Tianyuan Zhang, Yue Wang, Yilun Wang, and Hang Zhao. FUTR3D: A unified sensor fusion framework for 3d detection, March 2022. URL: https://arxiv.org/abs/2203.10642, arXiv:arXiv:2203.10642.

59. Pranav Singh Chib and Pravendra Singh. Recent advancements in end-to-end autonomous driving using deep learning: A survey, July 2023. URL: https://arxiv.org/abs/2307.04370, arXiv:arXiv:2307.04370.
60. François Chollet. The implausibility of intelligence explosion, November 2017. Accessed: 2025-01-08. URL: https://medium.com/@francois.chollet/the-impossibility-of-intelligence-explosion-5be4a9eda6ec.
61. François Chollet. On the measure of intelligence, November 2019. URL: https://arxiv.org/abs/1911.01547, arXiv:arXiv:1911.01547.
62. François Chollet. OpenAI o3 breakthrough high score on ARC-AGI-Pub, December 2024. Accessed: 2025-01-08. URL: https://arcprize.org/blog/oai-o3-pub-breakthrough.
63. Francois Chollet. Pattern recognition vs true intelligence. Machine Learning Street Talk, YouTube podcast, November 2024. URL: https://www.youtube.com/watch?v=JTU8Ha4Jyfc&t=3637s.
64. Sauhaarda Chowdhuri, Tushar Pankaj, and Karl Zipser. Multinet: Multi-modal multi-task learning for autonomous driving, September 2017. URL: https://arxiv.org/pdf/1709.05581, arXiv:arXiv:1709.05581.
65. Clayton M. Christensen, Michael E. Raynor, and Rory McDonald. What is disruptive innovation? *Harvard Business Review*, December 2015. Accessed: 2025-01-08. URL: https://hbr.org/2015/12/what-is-disruptive-innovation.
66. Alonzo Church. An unsolvable problem of elementary number theory. *American Journal of Mathematics*, 58(2):345–363, 1936. URL: https://ics.uci.edu/~lopes/teaching/inf212W12/readings/church.pdf.
67. Dan Claudiu Cireşan, Ueli Meier, Luca Maria Gambardella, and Jürgen Schmidhuber. Deep, big, simple neural nets for handwritten digit recognition. *Neural Computation*, 22(12):3207–3220, December 2010. URL: https://dl.acm.org/doi/abs/10.1162/NECO_a_00052, doi:10.1162/NECO_a_00052.
68. M. B. Clowes. On seeing things. *Artificial Intelligence*, 2:79–116, 1971. URL: https://www.sciencedirect.com/science/article/abs/pii/0004370271900051.
69. Jeremy Cohen. Why i believe Tesla still secretly uses CNNs in FSD12 (and not just transformers), 2024. Accessed: 2025-01-08. URL: https://www.thinkautonomous.ai/blog/tesla-cnns-vs-transformers/.
70. Jonathan H. Connell. Minimalist mobile robotics, 1990. URL: https://www.sciencedirect.com/book/9780121852306/minimalist-mobile-robotics.
71. Corinna Cortes and Vladimir Vapnik. Support-vector networks. *Machine Learning*, 20(3):273–297, 1995. URL: https://link.springer.com/article/10.1007/BF00994018, doi:10.1007/BF00994018.
72. Ian Craig and Paul Nisbet. The smart wheelchair: An augmentative mobility 'toolkit'. In *Proceedings of the 2nd European Conference on the Advancement of Rehabilitation Technology*, pages 131–134. Swedish Handicap Institute, Stockholm., May 1993.
73. Cristina Criddle and George Hammond. Openai bets on AI agents becoming mainstream by 2025, October 2024. Accessed March 2025. URL: https://www.ft.com/content/30677465-33bb-4f74-a8e6-239980091f7a.

74. George Cybenko. Approximations by superpositions of sigmoidal functions. *Mathematics of Control, Signals, and Systems*, 2(4):303–314, 1989. URL: https://link.springer.com/article/10.1007/BF02551274, doi:10.1007/BF02551274.
75. Erez Dagan. Solving the long-tail with E2E AI: 'the revolution will not be supervised', March 2024. Accessed: 2025-01-08. URL: https://wayve.ai/thinking/e2e-embodied-ai-solves-the-long-tail/.
76. J. L. Dallaway, R. D. Jackson, and P. H. A. Timmers. Rehabilitation robotics in europe. *IEEE Transactions on Rehabilitation Engineering*, 3(1):35–45, 1995. URL: https://ieeexplore.ieee.org/document/372890.
77. Alex Davies. *Driven: The Race to Create the Autonomous Car.* Simon and Schuster, 2021. URL: https://www.simonandschuster.com/books/Driven/Alex-Davies/9781501199455.
78. Nando de Freitas. The game is over! it's all about scale now! the secrets to AGI are within our reach., May 2022. URL: https://x.com/nandodf/status/1525397036325019649.
79. Pim de Haan, Dinesh Jayaraman, and Sergey Levine. Causal confusion in imitation learning, May 2019. URL: https://arxiv.org/abs/1905.11979, arXiv:arXiv:1905.11979.
80. Rina Dechter. Learning while searching in constraint-satisfaction problems. Technical Report CSD-860049 (R-58), University of California, Los Angeles, Computer Science Department, Cognitive Systems Laboratory, 1986. URL: https://ftp.cs.ucla.edu/tech-report/198_-reports/860049.pdf.
81. Defense Advanced Research Projects Agency. RACER revs up for checkered flag goal of high-speed, off-road autonomy, 2022. Accessed: 2025-01-02. URL: https://www.darpa.mil/news/2022/racer-off-road-autonomy.
82. Janosch Delcker. The man who invented the self-driving car (in 1986). *Politico*, July 2018. URL: https://www.politico.eu/article/delf-driving-car-born-1986-ernst-dickmanns-mercedes/#:~:text=Long%20before%20Big%20Tech%20got,onto%20the%20roads%20of%20Europe.
83. Daniel C. Dennett. Can machines think? with postscripts 1985 and 1997. In *Brainchildren: Essays on Designing Minds*, pages 3–29. MIT Press, Cambridge, MA, 1998. URL: https://mitpress.mit.edu/9780262540902/brainchildren/.
84. Arden Dertat. Applied deep learning - part 3: Autoencoders, October 2017. URL: https://medium.com/data-science/applied-deep-learning-part-3-autoencoders-1c083af4d798.
85. René Descartes. *Meditations on First Philosophy.* Cambridge Texts in the History of Philosophy. Cambridge University Press, Cambridge, 2 edition, 1641. Originally published in Latin as 'Meditationes de prima philosophia'. URL: https://www.cambridge.org/highereducation/books/descartes-meditations-on-first-philosophy/23589F72498100827E1B2674169C934#overview.
86. Jacob Devlin, Ming-Wei Chang, Kenton Lee, and Kristina Toutanova. BERT: Pre-training of deep bidirectional transformers for language understanding,

October 2018. URL: https://arxiv.org/abs/1810.04805, arXiv:arXiv:1810.04805.

87. Ernst Dickmanns. Dynamic vision as key element for AGI. After-dinner talk at the 4th Conference on Artificial General Intelligence (AGI-11), August 2011. URL: https://www.youtube.com/watch?v=YZ6nPhUG2i0.
88. Ernst Dickmanns. May a pair of 'eyes' be optimal for vehicles too? *Electronics*, 9(5), 2020. URL: https://www.mdpi.com/2079-9292/9/5/759, doi:10.3390/electronics9050759.
89. Ernst D. Dickmanns. Vehicles capable of dynamic vision: A new breed of technical beings? *Artificial Intelligence*, 103(1-2):49–76, 1998. URL: https://www.sciencedirect.com/science/article/pii/S000437029800071X, doi:10.1016/S0004-3702(98)00071-X.
90. Ernst D. Dickmanns. *Dynamic Vision for Perception and Control of Motion*. Springer, London, 2007. URL: https://link.springer.com/book/10.1007/978-1-84628-638-4, doi:10.1007/978-1-84628-638-4.
91. Ben Dickson. Tesla AI chief explains why self-driving cars don't need lidar, July 2021. Accessed: 2025-01-08. URL: https://venturebeat.com/business/tesla-ai-chief-explains-why-self-driving-cars-dont-need-lidar/.
92. G. P. Dinneen. Programming pattern recognition. In *Proceedings of the Western Joint Computer Conference (WJCC)*, pages 97–100, Los Angeles, CA, 1955. ACM/IEEE. URL: https://dl.acm.org/doi/10.1145/1455292.1455311.
93. Marco Dorigo, Vittorio Maniezzo, and Alberto Colorni. Positive feedback as a search strategy. In *Tech. Rep. 91-016*, Politecnico di Milano, Italy, 1991.
94. Alexey Dosovitskiy, Lucas Beyer, Alexander Kolesnikov, Dirk Weissenborn, Xiaohua Zhai, Thomas Unterthiner, Mostafa Dehghani, Matthias Minderer, Georg Heigold, Sylvain Gelly, Jakob Uszkoreit, and Neil Houlsby. An image is worth 16x16 words: Transformers for image recognition at scale, October 2020. URL: https://arxiv.org/abs/2010.11929, arXiv:arXiv:2010.11929.
95. Hubert L. Dreyfus and Stuart E. Dreyfus. Making a mind versus modelling the brain: Artificial intelligence back at the branchpoint. In Massimo Negrotti, editor, *Understanding the Artificial: On the Future Shape of Artificial Intelligence*, pages 33–54. Springer, London, 1991. doi:10.1007/978-1-4471-1776-6_3.
96. Harris Drucker, Christopher J. C. Burges, Linda Kaufman, Alex Smola, and Vladimir Vapnik. Support vector regression machines. In M. C. Mozer, M. Jordan, and T. Petsche, editors, *Advances in Neural Information Processing Systems*, volume 9, pages 155–161. MIT Press, 1996. URL: https://proceedings.neurips.cc/paper_files/paper/1996/file/d38901788c533e8286cb6400b40b386d-Paper.pdf.
97. Editorial. History of autonomous vehicles - timeline, March 17 2021. Accessed: 2025-01-03. URL: https://roboticsbiz.com/history-of-autonomous-vehicles-timeline/#google_vignette.
98. Encyclopædia Britannica. Nouvelle artificial intelligence. https://www.britannica.com/technology/nouvelle-artificial-intelligence, 2024. Accessed: 2025-02-12.

99. Encyclopaedia Britannica. Death in ancient Egypt, n.d. Accessed: 2025-01-08. URL: https://www.britannica.com/science/death/Ancient-Egypt.
100. EU. Technology initiative for the disabled and elderly: Project no. 1045 - SENARIO, 1994–1997. Bridge Phase, Budget: 997k ECU, Funding: 700k ECU. URL: https://op.europa.eu/en/publication-detail/-/publication/b36e36bf-c049-4162-8a22-124acec1cf30.
101. T. Evans. A heuristic program to solve geometric analogy problems. In *Proceedings of the Joint Computer Conference*, pages 327–338, April 1964. URL: https://dl.acm.org/doi/10.1145/1464122.1464156.
102. Edward A. Feigenbaum, Bruce G. Buchanan, and Joshua Lederberg. On generality and problem solving: A case study using the DENDRAL program. *Artificial Intelligence*, 2(1):1–28, 1971. URL: https://exhibits.stanford.edu/feigenbaum/catalog/zr438xj0026.
103. David Ferrucci, Eric Brown, Jennifer Chu-Carroll, James Fan, David Gondek, Aditya A. Kalyanpur, Adam Lally, J. William Murdock, Eric Nyberg, John Prager, Nico Schlaefer, and Chris Welty. Building Watson: An overview of the DeepQA project. *AI Magazine*, 31(3):59–79, July 2010. URL: https://ojs.aaai.org/aimagazine/index.php/aimagazine/article/view/2303, doi:10.1609/aimag.v31i3.2303.
104. Martin Ford. *Architects of Intelligence: The Truth About AI from the People Building It*. Packt Publishing, 2018. URL: https://www.packtpub.com/en-us/product/architects-of-intelligence-9781789959574.
105. Martin Ford. *Architects of Intelligence: The Truth About AI from the People Building It*. Packt Publishing, 2018. URL: https://www.packtpub.com/en-gb/product/architects-of-intelligence-9781789959574?srsltid=AfmBOop46eV1mL1DA-G1IRNlAqpJ7Z3ZQ-vEZfYmk4PLQssSOt1uY4pc.
106. Michel Foucault. *The Order of Things: An Archaeology of the Human Sciences*. Pantheon Books, New York, 1970. Originally published as *Les Mots et les Choses*, 1966, Éditions Gallimard, Paris. URL: https://www.penguinrandomhouse.com/books/55037/the-order-of-things-by-michel-foucault/.
107. Kunihiko Fukushima. Neocognitron: A self-organizing neural network model for a mechanism of pattern recognition unaffected by shift in position. *Biological Cybernetics*, 36(4):193–202, April 1980. URL: https://link.springer.com/article/10.1007/BF00344251.
108. Dennis Gabor. Theory of communication. *Journal of the Institution of Electrical Engineers - Part III: Radio and Communication Engineering*, 93(26):429–441, 1946. URL: https://jcsphysics.net/lit/gabor1946.pdf, doi:10.1049/ji-3-1.1946.0073.
109. Yarin Gal and Zoubin Ghahramani. Dropout as a Bayesian approximation: Representing model uncertainty in deep learning. In *Proceedings of the 33rd International Conference on Machine Learning (ICML)*, pages 1050–1059. JMLR.org, 2016. URL: https://proceedings.mlr.press/v48/gal16.html.
110. Roman Garnett. *Bayesian Optimization*. Cambridge University Press, 2023. URL: https://bayesoptbook.com/.

111. Tom M. Gasser, Clemens Arzt, Mihiar Ayoubi, Arne Bartels, Jana Eier, Frank Flemisch, Dirk Häcker, Tobias Hesse, Werner Huber, Christine Lotz, Markus Maurer, Simone Ruth-Schumacher, Jürgen Schwarz, and Wolfgang Vogt. Rechtsfolgen zunehmender fahrzeugautomatisierung. Technical Report Heft F83, Bundesanstalt für Straßenwesen, January 2012. URL: https://www.bast.de/DE/Publikationen/Berichte/unterreihe-f/2013-2012/f83.html.
112. Erann Gat. On three-layer architectures. In D. Kortenkamp, R. P. Bonnasso, and R. Murphy, editors, *Artificial Intelligence and Mobile Robotics*, pages 195–210. AAAI Press, 1998. URL: https://citeseerx.ist.psu.edu/document?repid=rep1&type=pdf&doi=f9e1f0b56761f07ac4ade715bcf9eb76e09de651.
113. Sean Gerrish. *How Smart Machines Think*. MIT Press, 2018. URL: https://mitpress.mit.edu/9780262537971/how-smart-machines-think/.
114. Ghost Autonomy. Self-driving for everyone. White paper, Ghost, 2023. Tech Brief, retrievable only via Wayback Machine. Accessed: 2025-01-08. URL: https://web.archive.org/web/20230520144453/https://assets.website-files.com/62deab46debedb5ae0390c30/640a484180fc6a98d8c029d9_Ghost-Autonomy-Engine-Tech-Brief.pdf.
115. Samuel Gibbs. Google almost bought Tesla when it had just two weeks of cash left, 2015. URL: https://www.theguardian.com/technology/2015/apr/21/google-almost-bought-tesla-elon-musk-larry-page.
116. Gerd Gigerenzer. *How to Stay Smart in a Smart World: Why Human Intelligence Still Beats Algorithms*. Penguin Books Ltd, 2022. URL: https://www.penguin.co.uk/books/321144/how-to-stay-smart-in-a-smart-world-by-gigerenzer-gerd/9780141995045.
117. Dayoung Gong, Joonseok Lee, Manjin Kim, Seong Jong Ha, and Minsu Cho. Future transformer for long-term action anticipation, May 2022. URL: https://arxiv.org/abs/2205.14022, arXiv:arXiv:2205.14022.
118. Ian Goodfellow, Jean Pouget-Abadie, Mehdi Mirza, Bing Xu, David Warde-Farley, Sherjil Ozair, Aaron Courville, and Yoshua Bengio. Generative adversarial nets. In *Advances in Neural Information Processing Systems*, volume 27, pages 2672–2680, 2014. URL: https://papers.nips.cc/paper/5423-generative-adversarial-nets.
119. Ian J. Goodfellow, David Warde-Farley, Mehdi Mirza, Aaron Courville, and Yoshua Bengio. Maxout networks. In *Proceedings of the 30th International Conference on Machine Learning*, pages 1319–1327, 2013. URL: https://proceedings.mlr.press/v28/goodfellow13.html.
120. Nicholas J. Gordon, David J. Salmond, and Andrew F. M. Smith. Novel approach to nonlinear and non-gaussian Bayesian state estimation. *IEE Proceedings F (Radar and Signal Processing)*, 140(2):107–113, 1993. URL: https://digital-library.theiet.org/doi/10.1049/ip-f-2.1993.0015, doi:10.1049/ip-f-2.1993.0015.
121. Rohan Goswami. Tesla's market cap is tied to solving autonomous driving, June 2023. Accessed: 2025-01-08. URL: https://www.cnbc.com/2023/06/16/elon-musk-teslas-market-cap-is-tied-to-solving-autonomous-driving.html.

122. Tristan Greene. Deepmind researcher claims new 'Gato' AI could lead to AGI, 2022. Accessed: 2025-01-08. URL: https://thenextweb.com/news/deepmind-researcher-claims-new-gato-ai-could-lead-to-agi-says-game-is-over.
123. Stephen Grossberg. Competitive learning: From interactive activation to adaptive resonance. *Cognitive Science*, 11(1):23–63, 1987. URL: https://onlinelibrary.wiley.com/doi/abs/10.1111/j.1551-6708.1987.tb00862.x, doi:10.1207/s15516709cog1101_2.
124. A. Guzzmann. Computer recognition of a visual field into three-dimensional bodies. In *AFIPS Conference Proceedings*, volume 33, pages 291–304, Washington, DC, 1968. URL: https://dspace.mit.edu/handle/1721.1/6902.
125. David Ha and Jürgen Schmidhuber. World models, March 2018. URL: https://arxiv.org/abs/1803.10122, arXiv:arXiv:1803.10122.
126. Alon Halevy, Peter Norvig, and Fernando Pereira. The unreasonable effectiveness of data. *IEEE Intelligent Systems*, pages 8–12, 2009. URL: https://ieeexplore.ieee.org/document/4804817.
127. Jiawei Han, Micheline Kamber, and Jian Pei. *Data Mining: Concepts and Techniques*. Morgan Kaufmann, Burlington, MA, 3rd edition, 2011. URL: https://shop.elsevier.com/books/data-mining/han/978-0-12-811760-6.
128. Karen Hao. The AI technique that could imbue machines with the ability to reason. *MIT Technology Review*, 2019. URL: https://www.technologyreview.com/2019/07/12/65579/the-next-ai-revolution-will-come-from-machine-learnings-most-underrated-form/.
129. Roy Harris. *Introduction to Integrationist Linguistics*. Pergamon Press, Oxford, UK, 1998.
130. W. K. Hastings. Monte Carlo sampling methods using markov chains and their applications. *Biometrika*, 57(1):97–109, 1970. URL: https://www.jstor.org/stable/2334940, doi:10.1093/biomet/57.1.97.
131. John Haugeland. *Artificial Intelligence: The Very Idea*. MIT Press, Cambridge, MA, 1985. URL: https://direct.mit.edu/books/book/4347/Artificial-IntelligenceThe-Very-Idea.
132. Andrew J. Hawkins. Meet ALVINN, the self-driving car from 1989, November 2016. Accessed: 2025-01-03. URL: https://www.theverge.com/2016/11/27/13752344/alvinn-self-driving-car-1989-cmu-navlab.
133. Andrew J. Hawkins. Tesla cybercab announced: Elon Musk's robotaxi is finally here, October 2024. Accessed: 2025-01-08. URL: https://www.theverge.com/2024/10/10/24265530/tesla-robotaxi-elon-musk-features-range-price-release-date.
134. John Hayes. A new way to drive: New learning systems will overtake robotics in the race to self-driving, January 2020. URL: https://web.archive.org/web/20221129015836/https://www.ghostautonomy.com/blog/a-new-way-to-drive.
135. Will Douglas Heaven. How Wayve's driverless cars will meet one of their biggest challenges yet, October 2024. Accessed: 2025-01-08. URL: https://www.technologyreview.com/2024/10/24/1106156/wayve-driverless-cars-us-biggest-challenges-yet/.

136. Richard Heimann and Clayton Pummill. *Generative Artificial Intelligence Revealed: Understanding AI from Technical Depth to Business Insights to Responsible Adoption.* Gracie House Publishing, 2023. Kindle Edition.
137. Alex Hern and Dan Milmo. Spam, junk ... slop? the latest wave of AI behind the 'zombie internet', May 2024. Accessed: 2025-05-14. URL: https://www.theguardian.com/technology/article/2024/may/19/spam-junk-slop-the-latest-wave-of-ai-behind-the-zombie-internet.
138. Alberto Hernández-Espinosa, Luan Ozelim, Felipe S. Abrahão, and Hector Zenil. Superarc: A test for general and super intelligence based on first principles of recursion theory and algorithmic probability, March 2025. URL: https://arxiv.org/abs/2503.16743, arXiv:arXiv:2503.16743.
139. Geoffrey E. Hinton, Simon Osindero, and Yee Whye Teh. A fast learning algorithm for deep belief nets. *Neural Computation*, 18(7):1527–1554, 2006. doi:10.1162/neco.2006.18.7.1527.
140. Geoffrey E. Hinton and Terry J. Sejnowski. Learning and relearning in Boltzmann machines. In *Proceedings of the 1985 Conference on Neural Information Processing Systems*, pages 358–365. IEEE, 1985. URL: https://www.cs.toronto.edu/~hinton/absps/pdp7.pdf.
141. Jerry Hirsch. Reality check: $160 billion can't get autonomous vehicles on the road, November 2022. Accessed: 2025-01-08. URL: https://www.autonews.com/mobility-report/autonomous-vehicle-reality-check-after-160-billion-spent/.
142. Jonathan Ho, Ajay Jain, and Pieter Abbeel. Denoising diffusion probabilistic models, June 2020. URL: https://arxiv.org/abs/2006.11239, arXiv:arXiv:2006.11239.
143. Thomas Hobbes. *Leviathan, or the Matter, Forme and Power of a Commonwealth Ecclesiastical and Civil.* Andrew Crooke, London, 1651. URL: https://quod.lib.umich.edu/cgi/t/text/text-idx?c=eebo;idno=A43998.0001.001.
144. Sepp Hochreiter and Jürgen Schmidhuber. Long short-term memory. *Neural Computation*, 9(8):1735–1780, 1997. URL: https://direct.mit.edu/neco/article-abstract/9/8/1735/6109/Long-Short-Term-Memory?redirectedFrom=fulltext.
145. A. L. Hodgkin and A. F. Huxley. A quantitative description of membrane current and its application to conduction and excitation in nerve. *Journal of Physiology*, 117(4):500–544, 1952. URL: https://physoc.onlinelibrary.wiley.com/doi/abs/10.1113/jphysiol.1952.sp004764.
146. Douglas R. Hofstadter. *Gödel, Escher, Bach: An Eternal Golden Braid.* Basic Books, New York, 1979. Winner of the 1980 Pulitzer Prize for General Nonfiction.
147. Jakob Hohwy. *The Predictive Mind.* Oxford University Press, Oxford, United Kingdom, 2013. URL: https://academic.oup.com/book/4105.
148. Toshiteru Homma, Les E. Atlas, and Robert J. Marks II. An artificial neural network for spatio-temporal bipolar patterns: Application to phoneme classification. In *Proceedings of the First Conference on Neural Information Processing*

Systems, pages 1–6. Neural Information Processing Systems Foundation, 1987. URL: https://www.researchgate.net/publication/221617843_An_Artificial_Neural_Network_for_Spatio-Temporal_Bipolar_Patterns_Application_to_Phoneme_Classification.

149. Sara Hooker. The hardware lottery, September 2020. URL: http://arxiv.org/pdf/2009.06489, arXiv:arXiv:2009.06484.
150. Joseph Hooper. From DARPA grand challenge 2004: DARPA's debacle in the desert, June 2004. URL: https://web.archive.org/web/20110702095855/http://www.popsci.com/scitech/article/2004-06/darpa-grand-challenge-2004darpas-debacle-desert.
151. John J. Hopfield. Neural networks and physical systems with emergent collective computational abilities. *Proceedings of the National Academy of Sciences*, 79(8):2554–2558, 1982. URL: https://www.pnas.org/doi/10.1073/pnas.79.8.2554.
152. Kurt Hornik, Maxwell Stinchcombe, and Halbert White. Multilayer feedforward networks are universal approximators. *Neural Networks*, 2(5):359–366, 1989. URL: https://www.sciencedirect.com/science/article/abs/pii/0893608089900208, doi:10.1016/0893-6080(89)90020-8.
153. Yiqi Hou, Sascha Hornauer, and Karl Zipser. Fast recurrent fully convolutional networks for direct perception in autonomous driving, November 2017. URL: https://arxiv.org/pdf/1711.06459, arXiv:arXiv:1711.06459.
154. H. Hoyer, U. Borgolte, and A. Jochheim. The OMNI-wheelchair - state of the art. In *CSUN 99: 14th "Technology and Persons with Disabilities" Conference*, Los Angeles, California, USA, March 1999. URL: https://www.dinf.ne.jp/doc/english/Us_Eu/conf/csun_99/session0274.html.
155. Jensen Huang. Accelerating AI with gpus: A new computing model, January 2016. URL: https://blogs.nvidia.com/blog/accelerating-ai-artificial-intelligence-gpus/.
156. David H. Hubel and Torsten N. Wiesel. Receptive fields, binocular interaction and functional architecture in the cat's visual cortex. *The Journal of Physiology*, 160(1):106–154, 1962. URL: https://pmc.ncbi.nlm.nih.gov/articles/PMC1359523/.
157. D. A. Huffman. Impossible objects as nonsense sentences. In B. Meltzer and D. Michie, editors, *Machine Intelligence 6*, pages 295–323. Wiley, New York, 1971.
158. Marcus Hutter, David Quarel, and Elliot Catt. *An Introduction to Universal Artificial Intelligence*. Chapman & Hall, London, 2024. URL: http://www.hutter1.net/publ/uaibook2.pdf.
159. Jyh-Jing Hwang, Runsheng Xu, Hubert Lin, Wei-Chih Hung, Jingwei Ji, Kristy Choi, Di Huang, Tong He, Paul Covington, Benjamin Sapp, Yin Zhou, James Guo, Dragomir Anguelov, and Mingxing Tan. EMMA: End-to-end multimodal model for autonomous driving, October 2024. URL: https://arxiv.org/abs/2410.23262, arXiv:arXiv:2410.23262.

160. AGI Sentinel Initiative. A working list: Definitions of artificial intelligence and human intelligence, 2020. URL: https://agisi.org/Defs_intelligence.html.
161. A. G. Ivakhnenko and V. G. Lapa. *Cybernetic Predicting Devices.* CCM Information Corporation, New York, 1965. Translated from Russian. URL: https://gwern.net/doc/ai/1966-ivakhnenko.pdf.
162. Vladimir G. Ivancevic and Tijana T. Ivancevic. *Computational Mind: A Complex Dynamics Perspective.* Springer, Berlin, Heidelberg, 2007. URL: https://link.springer.com/book/10.1007/978-3-540-71561-0.
163. D. N. Jagadish, A. Chauhan, and L. Mahto. Conditional variational autoencoder networks for autonomous vehicle path prediction. *Neural Processing Letters*, 54(5):3965–3978, 2022. doi:10.1007/s11063-022-10802-z.
164. Frederick Jelinek. Some of my best friends are linguists. *Language Resources and Evaluation*, 39(1):25–34, 2005. URL: https://www.researchgate.net/publication/220147830_Some_of_my_Best_Friends_are_Linguists.
165. Chiyu Max Jiang, Andre Cornman, Cheolho Park, Ben Sapp, Yin Zhou, and Dragomir Anguelov. MotionDiffuser: Controllable multi-agent motion prediction using diffusion, June 2023. 5th June, 2023. URL: https://arxiv.org/abs/2306.03083, arXiv:arXiv:2306.03083.
166. Cameron R. Jones and Benjamin K. Bergen. Does GPT-4 pass the Turing test?, October 2023. URL: https://arxiv.org/abs/2310.20216, arXiv:arXiv:2310.20216.
167. Nicola Jones. How should we test AI for human-level intelligence? OpenAI's o3 electrifies quest. *Nature*, 637:774–775, 2025. URL: https://www.nature.com/articles/d41586-025-00110-6, doi:10.1038/d41586-025-00110-6.
168. Michael I. Jordan, Zoubin Ghahramani, Tommi S. Jaakkola, and Lawrence K. Saul. An introduction to variational methods for graphical models. *Machine Learning*, 37(2):183–233, 1999. URL: https://link.springer.com/chapter/10.1007/978-94-011-5014-9_5, doi:10.1007/978-94-011-5014-9_5.
169. Kevin Jost. Ghost aims to democratize consumer car autonomy using generalized AI, August 2023. Accessed: 2025-01-08. URL: https://insideautonomousvehicles.com/ghost-aims-to-democratize-consumer-car-autonomy-using-generalized-ai/.
170. Jennifer Kahn. It's alive, March 2002. URL: https://www.wired.com/2002/03/everywhere.
171. R. E. Kalman. A new approach to linear filtering and prediction problems. *Journal of Basic Engineering*, 82(1):35–45, 1960. URL: https://asmedigitalcollection.asme.org/fluidsengineering/article-abstract/82/1/35/397706/A-New-Approach-to-Linear-Filtering-and-Prediction?redirectedFrom=fulltext.
172. L. N. Kanal and N. C. Randall. Recognition system design by statistical analysis. In *Proceedings of the 1964 19th ACM National Conference*, pages 196–202. Association for Computing Machinery, 1964. URL: https://dl.acm.org/doi/10.1145/800257.808903, doi:10.1145/800257.808903.
173. Laveen Kanal. Evaluation of a class of pattern-recognition networks. In Eugene E. Bernard and Morley R. Kare, editors, *Biological Prototypes and*

Synthetic Systems: Volume 1 Proceedings of the Second Annual Bionics Symposium sponsored by Cornell University and the General Electric Company, Advanced Electronics Center, held at Cornell University, August 30–September 1, 1961, pages 261–269. Springer US, Boston, MA, 1962. doi:10.1007/978-1-4684-1716-6_35.

174. Immanuel Kant. *Critique of Pure Reason*. Johann Friedrich Hartknoch, 1781. Translated by J. M. D. Meiklejohn. URL: https://www.gutenberg.org/files/4280/4280-h/4280-h.htm.
175. Jared Kaplan, Sam McCandlish, Tom Henighan, Tom B. Brown, Benjamin Chess, Rewon Child, Scott Gray, Alec Radford, Jeffrey Wu, and Dario Amodei. Scaling laws for neural language models, January 2020. URL: https://arxiv.org/abs/2001.08361, arXiv:arXiv:2001.08361.
176. Andrej Karpathy and Li Fei-Fei. Deep visual-semantic alignments for generating image descriptions, December 2014. URL: https://arxiv.org/abs/1412.2306, arXiv:arXiv:1412.2306.
177. N. Katevas, N. M. Sgouros, S. Tzafestas, G. Papakonstantinou, P. Beattie, J. M. Bishop, P. Tsanakas, and D. Koutsouris. The autonomous mobile robot SENARIO: A sensor-aided intelligent navigation for powered wheelchairs. *IEEE Robotics and Automation*, 4(4):60–70, 1997. URL: https://ieeexplore.ieee.org/document/637806.
178. Tasha Keeney and Alex Kendall. Autonomous vehicles powered by end-to-end deep learning, February 2021. URL: https://www.ark-invest.com/podcast/ep-90-autonomous-vehicles-wayve-ai.
179. Tasha Keeney, Sam Korus, and Daniel Maguire. ARK's expected value for Tesla in 2029: $2,600 per share, 2024. Accessed: 2025-01-08. URL: https://www.ark-invest.com/articles/valuation-models/arks-tesla-price-target-2029.
180. Alex Kendall. The road to embodied AI, May 2024. Accessed: 2025-01-08. URL: https://wayve.ai/thinking/road-to-embodied-ai/.
181. Prannay Khosla. One model to rule the road?, November 2023. Accessed: 2025-01-08. URL: https://web.archive.org/web/20240228063835/https://www.ghostautonomy.com/blog/one-model-to-rule-the-road.
182. Hyeongjoo Kim and Dieter Schönecker, editors. *Kant and Artificial Intelligence*. De Gruyter, Berlin/Boston, 2022. URL: https://www.degruyter.com/document/doi/10.1515/9783110706611/html, doi:10.1515/9783110706611.
183. Beth Kindig. AI spending to exceed a quarter trillion next year, November 2024. Accessed: 2025-01-08. URL: https://www.forbes.com/sites/bethkindig/2024/11/14/ai-spending-to-exceed-a-quarter-trillion-next-year/.
184. S. Kirkpatrick, C. D. Gelatt, and M. P. Vecchi. Optimization by simulated annealing. *Science*, 220(4598):671–680, 1983. URL: https://www.science.org/doi/10.1126/science.220.4598.671, doi:10.1126/science.220.4598.671.
185. S. C. Kleene. Representation of events in nerve nets and finite automata. Technical Report RM-704, The Rand Corporation, December 1951. URL: https://www.rand.org/content/dam/rand/pubs/research_memoranda/2008/RM704.pdf.

186. Teuvo Kohonen. Self-organized formation of topologically correct feature maps. *Biological Cybernetics*, 43(1):59–69, 1982. URL: https://link.springer.com/article/10.1007/BF00337288, doi:10.1007/BF00337288.
187. Daphne Koller and Nir Friedman. *Probabilistic Graphical Models: Principles and Techniques*. MIT Press, 2009. URL: https://mitpress.mit.edu/9780262013192/probabilistic-graphical-models/.
188. Phil Koopman. No, Mercedes-Benz will NOT take the blame for a Drive Pilot crash, September 2023. URL: https://safeautonomy.blogspot.com/2023/09/no-mercedes-benz-will-not-take-blame.html.
189. Kirsten Korosec. Openai-backed ghost autonomy shuts down, April 2024. Accessed: 2025-07-18. URL: https://techcrunch.com/2024/04/03/openai-backed-ghost-autonomy-shuts-down/.
190. Alex Krizhevsky, Ilya Sutskever, and Geoffrey E. Hinton. Imagenet classification with deep convolutional neural networks. In *Advances in Neural Information Processing Systems 25 (NIPS 2012)*, pages 1097–1105. Curran Associates, Inc., 2012. URL: https://papers.nips.cc/paper_files/paper/2012/hash/c399862d3b9d6b76c8436e924a68c45b-Abstract.html.
191. Benjamin Kuipers et al. Shakey: From conception to history. *AI Magazine*, 2017. URL: http://ai.stanford.edu/~nilsson/OnlinePubs-Nils/General%20Essays/Shakey-aimag-17.pdf.
192. Susumu Kuno. A system for transformational analysis. In *COLING'65: Proceedings of the 1965 Conference on Computational Linguistics*, pages 1–48, USA, 1965. Association for Computational Linguistics. doi:10.3115/990314.990325.
193. Gongjin Lan and Qi Hao. End-to-end planning of autonomous driving in industry and academia: 2022-2023, December 2023. URL: https://arxiv.org/abs/2401.08658, arXiv:arXiv:2401.08658.
194. E. H. Land and J. J. McCann. Lightness and retinex theory. *Journal of the Optical Society of America*, 61(1):1–11, 1971. URL: https://opg.optica.org/josa/ViewMedia.cfm?uri=josa-61-1-1&seq=0&guid=771bf5aa-7ea7-4342-94a6-ee15c9137064&html=true.
195. Anders Boesen Lindbo Larsen, Søren Kaae Sønderby, Hugo Larochelle, and Ole Winther. Autoencoding beyond pixels using a learned similarity metric. In *Proceedings of The 33rd International Conference on Machine Learning*, pages 1558–1566, 2016. URL: https://proceedings.mlr.press/v48/larsen16.pdf.
196. John Launchbury. A darpa perspective on artificial intelligence. DARPA Presentation on YouTube, 2017. URL: https://www.youtube.com/watch?v=-O01G3tSYpU.
197. Yann LeCun. Generalization and network design strategies. Technical Report CRG-TR-89-4, University of Toronto, Connectionist Research Group, June 1989. URL: https://masters.donntu.ru/2012/fknt/umiarov/library/lecun.pdf.
198. Yann LeCun. A path towards autonomous machine intelligence, June 2022. Position paper. URL: https://openreview.net/pdf?id=BZ5a1r-kVsf.

199. Yann LeCun and Yoshua Bengio. Scaling learning algorithms towards AI. In Léon Bottou, Olivier Chapelle, Dennis DeCoste, and Jason Weston, editors, *Large-Scale Kernel Machines*, pages 321–360. MIT Press, 2007. URL: https://direct.mit.edu/books/edited-volume/3172/chapter-abstract/88105/Scaling-Learning-Algorithms-toward-AI?redirectedFrom=fulltext.
200. Yann LeCun, Yoshua Bengio, and Geoffrey Hinton. Deep learning. *Nature*, 521:436–444, 2015. URL: https://www.nature.com/articles/nature14539, doi:10.1038/nature14539.
201. Yann LeCun, Bernhard E. Boser, John S. Denker, Donnie Henderson, Richard E. Howard, Wayne E. Hubbard, and Lawrence D. Jackel. Handwritten digit recognition with a back-propagation network. In *Neural Information Processing Systems*, pages 396–404, 1989. URL: https://proceedings.neurips.cc/paper/1989/file/53c3bce66e43be4f209556518c2fcb54-Paper.pdf.
202. Yann LeCun, Léon Bottou, Yoshua Bengio, and Patrick Haffner. Gradient-based learning applied to document recognition. *Proceedings of the IEEE*, 86(11):2278–2324, 1998. URL: https://doi.org/10.1109/5.726791.
203. Yann LeCun, Eric Cosatto, Jan Ben, Urs Muller, and Beat Flepp. End-to-end learning of vision-based obstacle avoidance for off-road robots. In *Presented at the Learning@Snowbird Workshop*, April 2004. URL: https://cs.nyu.edu/~yann/research/dave/.
204. Yann LeCun and Ishan Misra. Self-supervised learning: The dark matter of intelligence, March 2021. Accessed: 2025-01-02. URL: https://ai.meta.com/blog/self-supervised-learning-the-dark-matter-of-intelligence/.
205. Yann LeCun, Dava Newman, and Ina Turpen Fried. Debating technology, January 2025. Panel discussion. URL: https://www.weforum.org/meetings/world-economic-forum-annual-meeting-2025/sessions/debating-technology/.
206. Shane Legg and Marcus Hutter. A collection of definitions of intelligence, June 2007. URL: https://arxiv.org/abs/0706.3639, arXiv:arXiv:0706.3639.
207. Shane Legg and Marcus Hutter. Universal intelligence: A definition of machine intelligence. *Minds and Machines*, 17(4):391–444, 2007. URL: https://link.springer.com/article/10.1007/s11023-007-9079-x, doi:10.1007/s11023-007-9079-x.
208. Gottfried Wilhelm Leibniz. *Monadology*. Unknown, 1714. German, originally written in French as 'La Monadologie'. URL: https://www.gutenberg.org/ebooks/39441.
209. J. Y. Lettvin, H. R. Maturana, W. S. McCulloch, and W. H. Pitts. What the frog's eye tells the frog's brain. *Proceedings of the IRE*, 47(11):1940–1951, 1959. URL: https://ieeexplore.ieee.org/document/4065609, doi:10.1109/JRPROC.1959.287207.
210. Song Li, Yu-Qin Deng, Zhi-Ling Zhu, Hong-Li Hua, and Ze-Zhang Tao. A comprehensive review on radiomics and deep learning for nasopharyngeal carcinoma imaging. *Diagnostics*, 11(9):1523, 2021. doi:10.3390/diagnostics11091523.

211. James Lighthill. Artificial intelligence: A general survey. In *Artificial Intelligence: A Paper Symposium*. UK Science Research Council, 1973. URL: https://www.chilton-computing.org.uk/inf/literature/reports/lighthill_report/p001.htm.
212. R. Linkser. From basic network principles to neural architecture: emergence of spatial-opponent cells. *Proceedings of the National Academy of Sciences*, 83(19):7508–7512, 1986. URL: https://www.pnas.org/doi/abs/10.1073/pnas.83.19.7508, doi:10.1073/pnas.83.19.7508.
213. Seppo Linnainmaa. Algoritmin kumulatiivinen pyöristysvirhe yksittäisten pyöristysvirheiden taylor-kehitelmänä. Master's thesis, University of Helsinki, 1970. In Finnish. URL: https://people.idsia.ch/~juergen/linnainmaa1970thesis.pdf.
214. H. C. Longuet-Higgins. A computer algorithm for reconstructing a scene from two projections. *Nature*, 293:133–135, 1981. URL: https://www.nature.com/articles/293133a0, doi:10.1038/293133a0.
215. Wilhelm Loots. Tesla's humanlike AI vision system, July 2021. URL: https://www.pressreader.com/south-africa/driven/20210701/282634625627189?srsltid=AfmBOoqA8DsGevjEEc7ZY46mt7AlPn657dEbFy9EaOoS9WiMoNTaZ7d-.
216. Josh Lowensohn. For $10,000, your car can drive itself, June 2014. Accessed: 2025-01-08. URL: https://www.theverge.com/2014/6/23/5834604/cruise-trying-to-reinvent-cruise-control-to-steer-brake-and-actually-drive-you.
217. Jinxiong Lu, Shoaib Azam, Gokhan Alcan, and Ville Kyrki. Data-driven diffusion models for enhancing safety in autonomous vehicle traffic simulations, October 2024. URL: https://arxiv.org/abs/2410.04809, arXiv:arXiv:2410.04809.
218. Sheng Lu, Irina Bigoulaeva, Rachneet Sachdeva, Harish Tayyar Madabushi, and Iryna Gurevych. Are emergent abilities in large language models just in-context learning?, September 2023. URL: https://arxiv.org/abs/2309.01809, arXiv:arXiv:2309.01809.
219. R. L. Madarasz, H. C. Heiny, R. F. Cromp, and N. M. Mazur. The design of an autonomous vehicle for the disabled. *IEEE Journal of Robotics and Automation*, 2(3):117–126, 1986. URL: https://ieeexplore.ieee.org/document/1087052.
220. Alireza Makhzani, Jonathon Shlens, Navdeep Jaitly, Ian Goodfellow, and Brendan Frey. Adversarial autoencoders, November 2015. URL: https://arxiv.org/abs/1511.05644, arXiv:arXiv:1511.05644.
221. Jitendra Malik. Visual control of driving: A bit of history. Presented at the 2017 BEARS Symposium: Panel 1 - Self-Driving Cars: Is Autonomous Driving AI-Complete?, 2017. URL: https://www.youtube.com/watch?v=ltbufoOON8Y.
222. Gary Marcus. CONFIRMED: LLMs have indeed reached a point of diminishing returns, 2024. Accessed: 2025-01-08. URL: https://garymarcus.substack.com/p/confirmed-llms-have-indeed-reached.

223. David Marr. *Vision*. MIT Press, 1982. URL: https://mitpress.mit.edu/9780262514620/vision/.
224. David Marr and Ellen Hildreth. Theory of edge detection. *Proceedings of the Royal Society of London. Series B. Biological Sciences*, 207(1167):187–217, 1980. URL: https://royalsocietypublishing.org/doi/10.1098/rspb.1980.0020, doi:10.1098/rspb.1980.0020.
225. David Marr and Tomaso Poggio. From understanding computation to understanding neural circuitry. *Neurosciences Research Program Bulletin*, 15:470–488, 1977. URL: https://dspace.mit.edu/bitstream/handle/1721.1/5782/AIM-357.pdf.
226. Richard Mason, Jim Radford, Deepak Kumar, Robert Walters, Brian Fulkerson, Eagle Jones, David Caldwell, Jason Meltzer, Yaniv Alon, Amnon Shashua, Hiroshi Hattori, Emilio Frazzoli, and Stefano Soatto. The Golem Group / UCLA Autonomous ground vehicle in the DARPA Grand Challenge. *Journal of Field Robotics*, 23(8):527–553, 2006. URL: https://onlinelibrary.wiley.com/doi/10.1002/rob.20137, doi:10.1002/rob.20137.
227. C. Matias and J. Chen. Judging creativity in AI-generated art: Chinese and western perspectives on originality in copyright law. *International Journal of Legal Discourse*, 9(2):339–365, 2024. doi:10.1515/ijld-2024-2015.
228. Fulvio Mazzocchi. Could big data be the end of theory in science?: A few remarks on the epistemology of datadriven science. *EMBO Reports*, 16(10):1250–1255, 2015. URL: https://pmc.ncbi.nlm.nih.gov/articles/PMC4766450/.
229. J. McCarthy and P.J. Hayes. Some philosophical problems from the standpoint of artificial intelligence. In Bonnie Lynn Webber and Nils J. Nilsson, editors, *Readings in Artificial Intelligence*, pages 431–450. Morgan Kaufmann, 1981. URL: https://www.sciencedirect.com/science/article/pii/B9780934613033500337, doi:10.1016/B978-0-934613-03-3.50033-7.
230. John McCarthy. Computer controlled cars, 1968. Accessed: 2025-07-16. URL: http://jmc.stanford.edu/commentary/progress/car.html.
231. John McCarthy. Ascribing mental qualities to machines. Technical Report CA 94305, Stanford University, 1979. URL: https://www-formal.stanford.edu/jmc/ascribing.pdf.
232. John McCarthy. From here to human-level AI. *Artificial Intelligence*, 171(18):1174–1182, 2007. URL: https://linkinghub.elsevier.com/retrieve/pii/S0004370207001476.
233. John McCarthy. What is artificial intelligence? Technical Report CA 94305, Stanford University, 2007. URL: https://www-formal.stanford.edu/jmc/whatisai.pdf.
234. John McCarthy, Marvin L. Minsky, Nathaniel Rochester, and Claude E. Shannon. A proposal for the dartmouth summer research project on artificial intelligence, august 31, 1955. *AI Magazine*, 27(4), December 2006. URL: https://ojs.aaai.org/aimagazine/index.php/aimagazine/article/view/1904, doi:10.1609/aimag.v27i4.1904.

235. Pamela McCorduck. *Machines Who Think (2nd edition).* Routledge, 2004. URL: https://www.routledge.com/Machines-Who-Think-A-Personal-Inquiry-into-the-History-and-Prospects-of-Artificial-Intelligence/McCorduck/p/book/9781568812052?srsltid=AfmBOoqiKAOl9xh9oYfLQWqK7-4r3nX82wJV0r0yjWkq-nxgOsecLacu.
236. Warren McCulloch and Walter Pitts. A logical calculus of the ideas immanent in nervous activity. *Bulletin of Mathematical Biophysics*, 5:115–133, 1943. URL: https://link.springer.com/article/10.1007/BF02478259#citeas.
237. Michael E. McGrath. *Autonomous Vehicles: Opportunities, Strategies and Disruptions.* Independent Publishing, 2019.
238. Dan McLaughlin. America's first freeway: The Arroyo Seco Parkway, aka, the 110, April 2018. Accessed: 2025-01-08. URL: https://www.motortrend.com/features/1804-americas-first-freeway-the-arroyo-seco-parkway-aka-the-110/.
239. Chris S. Mellish, editor. *International Joint Conference on Artificial Intelligence IJCAI'95*, San Francisco, CA, USA, August 1995. Morgan Kaufmann Publishers Inc. Palais de Congres, Montreal, Quebec, Canada. August 20th–25th, 1995. URL: https://www.ijcai.org/proceedings/1995-1.
240. Luigi Federico Menabrea. Notions sur la machine analytique de M. Charles Babbage. *Bibliothèque universelle de Genève, nouvelle série*, 41, 1842. URL: https://archive.org/details/TO01056056.
241. Luigi Federico Menabrea. Sketch of the analytical engine invented by Charles Babbage, esq. *Scientific Memoirs*, 3, 1843. Originally published in French in 1842; translated and annotated by Ada Lovelace. URL: https://www.gutenberg.org/ebooks/75107.
242. K. De Meyer, J. M. Bishop, and S. J. Nasuto. Attention through self-synchronisation in the spiking neuron stochastic diffusion network. In *Proceedings of the 4th Association for the Scientific Study of Consciousness (ASSC4)*, 2000. URL: https://www.academia.edu/4093185/Attention_through_Self_Synchronisation_in_the_Spiking_Neuron_Stochastic_Diusion_Network.
243. Marvin Minsky. *Theory of Neural-Analogue Reinforcement Systems and its Application to the Brain Model Problem.* PhD thesis, Princeton University, Princeton, USA, 1954. URL: https://people.csail.mit.edu/brooks/idocs/Marvin_Minsky_PhD_Princeton_1953_1954.pdf.
244. Marvin Minsky and Seymour Papert. *Perceptrons: An Introduction to Computational Geometry.* MIT Press, Cambridge, MA, 1969. URL: https://mitpress.mit.edu/9780262630221/perceptrons/.
245. Marvin L. Minsky. *Computation: Finite and Infinite Machines.* Prentice-Hall, Inc., 1967. URL: https://dl.acm.org/doi/book/10.5555/1095587.
246. Melanie Mitchell and David C. Krakauer. The debate over understanding in ai's large language models. arXiv: [cs.LG], February 2023. 10th February 2023. URL: https://arxiv.org/pdf/2210.13966, arXiv:2210.13966.
247. Dagmar Monett and Colin W. P. Lewis. Getting clarity by defining artificial intelligence—a survey. In Vincent C. Müller, editor, *Philosophy and Theory of Artificial Intelligence 2017*, pages 212–214. Springer International Publishing,

2018. URL: https://link.springer.com/chapter/10.1007/978-3-319-96448-5_21#citeas.

248. Dagmar Monett, Colin W. P. Lewis, and Kristinn R. Thórisson, editors. *SOn Defining Artificial Intelligence—Commentaries and Author's Response*, volume 11:2 of *Journal of Artificial General Intelligence: special issue*. Walter de Gruyter, Inc., Boston, MA, USA, February 2020. URL: https://sciendo.com/article/10.2478/jagi-2020-0003.
249. Michael Montemerlo, Jan Becker, Suhrid Bhat, Hendrik Dahlkamp, Dmitri Dolgov, Scott Ettinger, Dirk Haehnel, Tim Hilden, Gabe Hoffmann, Burkhard Huhnke, Doug Johnston, Stefan Klumpp, Dirk Langer, Anthony Levandowski, Jesse Levinson, Julien Marcil, David Orenstein, Johannes Paefgen, Isaac Penny, Anna Petrovskaya, Mike Pflueger, Ganymed Stanek, David Stavens, Antone Vogt, and Sebastian Thrun. Junior: The stanford entry in the urban challenge. In *The third DARPA Urban Challenge: Autonomous Vehicles in City Traffic*, pages 91–123. Springer, 2009. URL: https://link.springer.com/chapter/10.1007/978-3-642-03991-1_3, doi:10.1007/978-3-540-78912-7_1.
250. J.H. Moor. The nature, importance, and difficulty of machine ethics. *IEEE Intelligent Systems*, 21(4):18–21, 2006. URL: https://ieeexplore.ieee.org/document/1667948, doi:10.1109/MIS.2006.80.
251. Meredith Ringel Morris, Jascha Sohl-dickstein, Noah Fiedel, Tris Warkentin, Allan Dafoe, Aleksandra Faust, Clement Farabet, and Shane Legg. Levels of AGI for operationalizing progress on the path to AGI, November 2023. URL: https://arxiv.org/abs/2311.02462, arXiv:arXiv:2311.02462.
252. Danielle Muoio. Tesla has settled a lawsuit against its former autopilot director, April 2017. URL: https://www.businessinsider.com/tesla-sterling-anderson-lawsuit-dropped-2017-4?r=US&IR=T.
253. Elon Musk, April 2021. URL: https://x.com/elonmusk/status/1382549229726498821?lang=de.
254. Elon Musk. Elon Musk discusses the war in Ukraine and the importance of nuclear power, March 2022. Accessed: 2025-01-08. URL: https://www.businessinsider.com/elon-musk-interview-axel-springer-tesla-war-in-ukraine-2022-3.
255. Elon Musk. Diffusion seems to be more compute-efficient than transformers for vision, April 2023. Accessed: 2025-01-08. URL: https://tinyurl.com/4xenkuf7.
256. Slawomir J. Nasuto, J. Mark Bishop, and Kurt De Meyer. Communicating neurons: A connectionist spiking neuron implementation of stochastic diffusion search. *Neurocomputing*, 72(4–6):704–712, 2009. URL: https://www.sciencedirect.com/science/article/abs/pii/S0925231208004608.
257. Slawomir J. Nasuto and John Mark Bishop. Of (zombie) mice and animats. In Vincent C. Müller, editor, *Philosophy and Theory of Artificial Intelligence*, pages 85–106. Springer Berlin Heidelberg, Berlin, Heidelberg, 2013. doi:10.1007/978-3-642-31674-6_7.

258. National Highway Traffic Safety Administration. Preliminary statement of policy concerning automated vehicles. Technical report, U.S. Department of Transportation, 2013. URL: https://www.nhtsa.gov/sites/nhtsa.gov/files/documents/automated_vehicles_policy.pdf.
259. National Highway Traffic Safety Administration. DOT/NHTSA policy statement concerning automated vehicles: 2016 update to "preliminary statement of policy concerning automated vehicles". Technical report, U.S. Department of Transportation, 2016. URL: https://www.nhtsa.gov/sites/nhtsa.gov/files/autonomous-vehicles-policy-update-2016.pdf.
260. John Von Neumann and Arthur W. Burks. *Theory of Self-Reproducing Automata*. University of Illinois Press, USA, 1966. URL: https://dl.acm.org/doi/10.5555/1102024.
261. A. Newell and H. Simon. The logic theory machine–a complex information processing system. *IRE Transactions on Information Theory*, 2(3):61–79, 1956. URL: https://ieeexplore.ieee.org/document/1056797, doi:10.1109/TIT.1956.1056797.
262. Allen Newell, J. C. Shaw, and Herbert A. Simon. GPS, a program that simulates human thought. In H. Billing, editor, *Lernende Automaten*, pages 109–124. Oldenbourg, Munich, 1960. URL: https://iiif.library.cmu.edu/file/Simon_box00064_fld04907_bdl0001_doc0001/Simon_box00064_fld04907_bdl0001_doc0001.pdf.
263. Allen Newell and Herbert A. Simon. Computer science as empirical inquiry: Symbols and search. *Communications of the ACM*, 19(3):113–126, 1976. URL: https://dl.acm.org/doi/10.1145/360018.360022.
264. Alva Nöe. Rage against the machine, October 2024. Accessed: 2025-01-02. URL: https://aeon.co/essays/can-computers-think-no-they-cant-actually-do-anything.
265. Nuro. CIML: Combining imitation and reinforcement learning for safe autonomous driving, July 2024. July 12th. URL: https://medium.com/nuro/cimrl-combining-imitation-and-reinforcement-learning-for-safe-autonomous-driving-13148ac99527.
266. Kyoung-Su Oh and Keechul Jung. GPU implementation of neural networks. *Pattern Recognition*, 37(6):1311–1314, 2004. URL: https://www.sciencedirect.com/science/article/abs/pii/S0031320304000524, doi:10.1016/j.patcog.2004.01.013.
267. Alan Ohnsman. At $1.1 billion Google's self-driving car moonshot looks like a bargain, 2017. URL: https://www.forbes.com/sites/alanohnsman/2017/09/15/at-1-1-billion-googles-self-driving-car-moonshot-looks-like-a-bargain/?sh=8c305c957bbf.
268. Paravisionlab. How to use LeNet5 and Keras for amazing MNIST results, 2024. URL: https://paravisionlab.co.in/lenet5-keras/.
269. Judea Pearl. Reverend Bayes on inference engines: A distributed hierarchical approach. In *Proceedings AAAI National Conference on AI*, pages 133–136, 1982. URL: https://ftp.cs.ucla.edu/pub/stat_ser/r30-reprint.pdf.

270. Mike Pesarchick. With self-driving tech was new, a road trip paved the way. *Associated Press*, August 2020. August 11th. URL: https://apnews.com/general-news-7eaf98e853534958a422ec71e100c4c0.
271. Tharin Pillay. How OpenAI's Sam Altman is thinking about AGI and superintelligence in 2025, January 2025. Accessed: 2025-01-10. URL: https://time.com/7205596/sam-altman-superintelligence-agi/.
272. Philip Polack, Florent Altché, Brigitte d'Andréa Novel, and Arnaud de La Fortelle. The kinematic bicycle model: A consistent model for planning feasible trajectories for autonomous vehicles? In *2017 IEEE Intelligent Vehicles Symposium (IV)*, pages 812–818. IEEE Press, 2017. doi:10.1109/IVS.2017.7995816.
273. Michael Poli, Stefano Massaroli, Eric Nguyen, Daniel Y. Fu, Tri Dao, Stephen Baccus, Yoshua Bengio, Stefano Ermon, and Christopher Ré. Hyena hierarchy: Towards larger convolutional language models, February 2023. 21st February. URL: https://arxiv.org/abs/2302.10866, arXiv:arXiv:2302.10866.
274. D. Pomerleau. Ralph: rapidly adapting lateral position handler. In *Proceedings of the Intelligent Vehicles Symposium IVS'95*, pages 506–511, 1995. URL: https://ieeexplore.ieee.org/document/528333, doi:10.1109/IVS.1995.528333.
275. Dean A. Pomerleau. Alvinn: an autonomous land vehicle in a neural network. In *Proceedings of the 2nd International Conference on Neural Information Processing Systems*, NIPS'88, pages 305–313, Cambridge, MA, USA, 1988. MIT Press. URL: https://dl.acm.org/doi/10.5555/2969735.2969771.
276. R. F. Port and T. van Gelder. *Mind as Motion: Explorations in the Dynamics of Cognition*. The MIT Press, Cambridge, MA, 1998. URL: https://direct.mit.edu/books/book/4292/Mind-as-MotionExplorations-in-the-Dynamics-of.
277. M. Ross Quillian. *Semantic Memory*, volume 66 (Issue 189) of *AFCRL*. Air Force Cambridge Research Laboratories, Office of Aerospace Research, United States Air Force, 1966.
278. Randolph Quirk, Sidney Greenbaum, Geoffrey Leech, and Jan Svartvik. *A Comprehensive Grammar of the English Language*. Longman, 1985. URL: https://openlibrary.org/books/OL2864840M/A_Comprehensive_grammar_of_the_English_language.
279. Rajat Raina, Anand Madhavan, and Andrew Ng. Large-scale deep unsupervised learning using graphics processors. In *ICML'09: Proceedings of the 26th International Conference on Machine Learning*, pages 873–880. Association for Computing Machinery, 2009. URL: https://dl.acm.org/doi/10.1145/1553374.1553486.
280. Reuters. Apple SVP says car is 'ultimate mobile device', May 2015. Accessed: 2025-01-08. URL: https://www.reuters.com/article/business/apple-svp-says-car-is-ultimate-mobile-device-idUSL1N0YI1OM/.
281. L.G. Roberts. *Machine Perception of Three-Dimensional Solids*. PhD thesis, Massachusetts Institute of Technology, Cambridge, Mass, 1963. URL: https://dspace.mit.edu/handle/1721.1/11589.

282. Herbert L. Roitblat. *Algorithms Are Not Enough: Creating General Artificial Intelligence*. MIT Press, 2020. URL: https://mitpress.mit.edu/9780262044127/algorithms-are-not-enough/.
283. Roohullah, Fazli Wahid, Sikandar Ali, Irshad Ahmed Abbasi, Samad Baseer, and Habib Ullah Khan. Accident detection in autonomous vehicles using modified restricted boltzmann machine. *Security and Communication Networks*, 2022(1):6424835, 2022. URL: https://onlinelibrary.wiley.com/doi/full/10.1155/2022/6424835, doi:10.1155/2022/6424835.
284. Robert Rosen. *Anticipatory Systems: Philosophical, Mathematical, and Methodological Foundations*. Pergamon Press, New York, 1985. URL: https://www.sciencedirect.com/book/9780080311586/anticipatory-systems.
285. Frank Rosenblatt. Two theorems of statistical separability in the perceptron. In *The Mechanisation of Thought Processes: Proceedings of a Symposium held at the National Physical Laboratory*, volume 1, pages 419–449. HMSO, London, 1958.
286. Frank Rosenblatt. *Principles of Neurodynamics: Perceptrons and the Theory of Brain Mechanisms*. Spartan Books, New York, 1962.
287. David E. Rumelhart, Geoffrey E. Hinton, and Ronald J. Williams. Learning representations by back-propagating errors. *Nature*, 323:533–536, 1986. URL: https://www.nature.com/articles/323533a0, doi:10.1038/323533a0.
288. Stuart J. Russell and Peter Norvig. *Artificial Intelligence: A Modern Approach*. Pearson, 4th edition, 2020. URL: https://aima.cs.berkeley.edu/.
289. Walid Saba. Deep misconceptions and the myth of data-driven language, 2021. URL: https://www.linkedin.com/pulse/deep-misconceptions-myth-data-driven-language-walid-saba.
290. Nikita Sachdeva. Understanding diffusion models: Types, real-world uses, and limitations. daffodil online article, August 2023. 21st August 2023. URL: https://insights.daffodilsw.com/blog/all-you-need-to-know-about-diffusion-models.
291. SAE International. Taxonomy and definitions for terms related to on-road motor vehicle automated driving systems (j3016), 2014. URL: https://www.sae.org/standards/content/j3016_201401/.
292. SAE International. Taxonomy and definitions for terms related to driving automation systems for on-road motor vehicles (j3016_201806), 2018. URL: https://www.sae.org/standards/content/j3016_201806/.
293. SAE International. Taxonomy and definitions for terms related to driving automation systems for on-road motor vehicles (j3016_201906), 2019.
294. SAE International. SAE levels of driving automation refined for clarity and international audience, 2021. URL: https://www.sae.org/blog/sae-j3016-update.
295. Sumit Saha. A comprehensive guide to convolutional neural networks — the ELI5 way. Medium article, December 2018. 15th December 2018. URL: https://medium.com/data-science/a-comprehensive-guide-to-convolutional-neural-networks-the-eli5-way-3bd2b1164a53.

296. A. L. Samuel. Some studies in machine learning using the game of checkers. *IBM Journal of Research and Development*, 3(3):210–229, 1959. URL: https://ieeexplore.ieee.org/document/5392560, doi:10.1147/rd.33.0210.
297. Arjun Sarkar. All you need to know about 'attention' and 'transformers' - in-depth understanding - part 1. Medium, February 2022. 15th February 2022. URL: https://medium.com/data-science/all-you-need-to-know-about-attention-and-transformers-in-depth-understanding-part-1-552f0b41d021.
298. Rylan Schaeffer, Brando Miranda, and Sanmi Koyejo. Are emergent abilities of large language models a mirage?, April 2023. 28th April 2023. URL: https://arxiv.org/abs/2304.15004, arXiv:arXiv:2304.15004.
299. Roger C. Schank. *Conceptual Information Processing*, volume 3 of *Fundamental Studies in Computer Science*. North-Holland Publishing Company, Amsterdam, 1975. URL: https://www.sciencedirect.com/book/9781483229737/conceptual-information-processing#book-description.
300. Michael Schick. Wozniak: Could a computer make a cup of coffee? Fast Company, March 2010. 2nd March 2010. URL: https://www.fastcompany.com/1568187/wozniak-could-computer-make-cup-coffee.
301. Jürgen Schmidhuber. 2011: First superhuman visual pattern recognition. AI blog, 2021. Originally published in 2011; minor updates for 10-year anniversary 2021. URL: https://people.idsia.ch/~juergen/superhumanpatternrecognition.html.
302. Jürgen Schmidhuber. Making the world differentiable: On using self-supervised fully recurrent neural networks for dynamic reinforcement learning and planning in non-stationary environments. Technical Report FKI-126-90 (revised), Technische Universität München, Institut für Informatik, 1990. URL: https://people.idsia.ch/~juergen/FKI-126-90_%28revised%29bw_ocr.pdf.
303. Jürgen Schmidhuber. Learning to control fast-weight memories: An alternative to recurrent nets. Technical Report FKI-147-91, Institut für Informatik, Technische Universität München, March 1991. URL: https://people.idsia.ch/~juergen/FKI-147-91ocr.pdf.
304. Jürgen Schmidhuber. Neural sequence chunkers. Technical Report FKI-148-91, Institut für Informatik, Technische Universität München, April 1991. URL: https://people.idsia.ch/~juergen/FKI-148-91ocr.pdf.
305. Jürgen Schmidhuber. A possibility for implementing curiosity and boredom in model-building neural controllers. In Jean-Arcady Meyer and Stewart W. Wilson, editors, *Proceedings of the International Conference on Simulation of Adaptive Behavior: From Animals to Animats*, pages 222–227. MIT Press/Bradford Books, 1991. NB. The IEEE site linked here egregiously credits the volume editors as the authors of this paper by Schmidhuber. URL: https://ieeexplore.ieee.org/document/6294131.
306. Jürgen Schmidhuber. Learning to control fast-weight memories: An alternative to dynamic recurrent networks. *Neural Computation*, 4(1):131–139, 1992. URL: https://doi.org/10.1162/neco.1992.4.1.131.
307. Jürgen Schmidhuber. Generative adversarial networks are special cases of artificial curiosity (1990) and also closely related to predictability minimization

(1991). *Neural Networks*, 127:58–66, 2020. URL: https://www.sciencedirect.com/science/article/abs/pii/S0893608020301283.

308. Jürgen Schmidhuber. Annotated history of modern AI and deep learning, December 2022. 21st December 2022. URL: https://arxiv.org/abs/2212.11279, arXiv:arXiv:2212.11279.
309. Jürgen Schmidhuber. Deep learning: Our miraculous year 1990-1991, May 2022. 12th May 2022. URL: https://arxiv.org/abs/2005.05744, arXiv:arXiv:2005.05744.
310. Jürgen Schmidhuber, Dan C. Cireşan, Ueli Meier, Jonathan Masci, and Alex Graves. On fast deep nets for AGI vision. In *Artificial General Intelligence: 4th International Conference, AGI 2011, Mountain View, CA, USA, August 3-6, 2011. Proceedings*, volume 6830 of *Lecture Notes in Computer Science*, pages 243–246, Berlin, Heidelberg, 2011. Springer. URL: https://link.springer.com/chapter/10.1007/978-3-642-22887-2_25, doi:10.1007/978-3-642-22887-2_25.
311. Jürgen Schmidhuber, Kristinn R. Thórisson, and Moshe Looks, editors. *Artificial General Intelligence: Proceedings of the 4th International Conference, AGI 2011*, volume 6830 of *Lecture Notes in Computer Science*, Berlin, Heidelberg, 2011. Springer. Mountain View, CA, USA. August 3rd-6th, 2011. URL: https://link.springer.com/book/10.1007/978-3-642-22887-2, doi:10.1007/978-3-642-22887-2.
312. Dan Schmidt, Nick Moran, Jonathan S. Rosenfeld, Jonathan Rosenthal, and Jonathan Yedidia. Self-play learning without a reward metric, December 2019. 16th December 2019. URL: https://arxiv.org/abs/1912.07557, arXiv:arXiv:1912.07557.
313. Maximilian Schreiner. Sam Altman on agi: Scaling large language models is not enough, November 2023. URL: https://the-decoder.com/sam-altman-on-agi-scaling-large-language-models-is-not-enough/.
314. John R. Searle. Minds, brains, and programs. *Behavioral and Brain Sciences*, 3(3):417–424, 1980. URL: https://www.cambridge.org/core/journals/behavioral-and-brain-sciences/article/abs/minds-brains-and-programs/DC644B47A4299C637C89772FACC2706A#, doi:10.1017/S0140525X00005756.
315. Oliver Selfridge. Pandemonium: A paradigm for learning. In *Proceedings of the Symposium on the Mechanisation of Thought Processes*, pages 513–526, London: HMSO, 1958. URL: https://aitopics.org/download/classics:504E1BAC.
316. Oliver Selfridge. Pattern recognition and modern computers. In *Proceedings of the March 1-3, 1955, Western Joint Computer Conference*, AFIPS'55 (Western), pages 91–93, New York, NY, USA, 1955. Association for Computing Machinery. URL: https://dl.acm.org/doi/10.1145/1455292.1455310, doi:10.1145/1455292.1455310.
317. Nikitas M. Sgouros. Qualitative navigation for autonomous wheelchair robots in indoor environments. *Autonomous Robots*, 12:257–266, 2002. URL: https://link.springer.com/article/10.1023/A:1015265514820, doi:10.1023/A:1015265514820.

318. Huma Shah and Kevin Warwick. Testing Turing's five minutes, parallelpaired imitation game. *Kybernetes*, 39(3):449–465, 2010. URL: https://www.emerald.com/insight/content/doi/10.1108/03684921011036178/full/html, doi:10.1108/03684921011036178.
319. Shai Shalev-Shwartz, Shaked Shammah, and Amnon Shashua. On a formal model of safe and scalable self-driving cars, August 2017. 21st August 2017. URL: https://arxiv.org/abs/1708.06374, arXiv:arXiv:1708.06374.
320. Amnon Shashua and Shai Shalev-Shwartz. Defining a new taxonomy for consumer autonomous vehicles. Mobileye Blog, February 2023. 6th February 2023. URL: https://www.mobileye.com/opinion/defining-a-new-taxonomy-for-consumer-autonomous-vehicles/.
321. David Sherrington and Scott Kirkpatrick. Solvable model of a spin-glass. *Physical Review Letters*, 35(26):1792–1796, 1975. doi:10.1103/PhysRevLett.35.1792.
322. David Silver, Aja Huang, Chris J. Maddison, Arthur Guez, Laurent Sifre, George van den Driessche, Julian Schrittwieser, Ioannis Antonoglou, Veda Panneershelvam, Marc Lanctot, Sander Dieleman, Dominik Grewe, John Nham, Nal Kalchbrenner, Ilya Sutskever, Timothy Lillicrap, Madeleine Leach, Koray Kavukcuoglu, Thore Graepel, and Demis Hassabis. Mastering the game of Go with deep neural networks and tree search. *Nature*, 529:484–489, January 2016. URL: https://www.nature.com/articles/nature16961, doi:10.1038/nature16961.
323. Nate Silver. *The Signal and the Noise: The Art and Science of Prediction.* Penguin Books Ltd, kindle edition edition, 2012. URL: https://www.penguin.co.uk/books/195518/the-signal-and-the-noise-by-silver-nate/9780141975658.
324. Gordon Bell. Architects look to processors of future. *Microprocessor Report*, 10(10):1–8, August 1996. 5th August 1996. URL: http://cva.stanford.edu/classes/cs99s/papers/architects_look_to_future.pdf.
325. Steven Skiena. *Calculated Bets: Computers, Gambling, and Mathematical Modeling to Win!* Cambridge University Press, August 2001. URL: https://www.cambridge.org/us/universitypress/subjects/computer-science/computing-general-interest/calculated-bets-computers-gambling-and-mathematical-modeling-win?format=PB&isbn=9780521009621.
326. B. F. Skinner. *The Behavior of Organisms: An Experimental Analysis.* Appleton-Century, New York, 1938. URL: https://psycnet.apa.org/record/1939-00056-000.
327. Adam Smith. Tesla's future 'worth basically zero' without self-driving cars. *The Independent*, June 2022. 16th June 2022. URL: https://www.independent.co.uk/tech/elon-musk-tesla-self-driving-b2102597.html.
328. Bradney Smith. Llms from scratch series — a complete guide to understanding and building large language models, December 2023. 11th December 2023. URL: https://medium.com/p/cedc9f72de4e.

329. Paul Smolensky. Information processing in dynamical systems: Foundations of harmony theory. In *Proceedings of the 1986 Conference on Neural Information Processing Systems*, pages 194–200, 1986. NB. The IEEE site linked here egregiously credits the volume editors as the authors of this paper by Smolensky. URL: https://ieeexplore.ieee.org/document/6302931/authors#authors.
330. Noah Snavely, Steven M. Seitz, and Richard Szeliski. Photo tourism: Exploring photo collections in 3D. *ACM Transactions on Graphics*, 25(3):835–846, 2006. URL: https://dl.acm.org/doi/10.1145/1141911.1141964, doi:10.1145/1141911.1141964.
331. Mathew Spencer, Etienne Roesch, Slawomir J. Nasuto, Thomas Tanay, and J. Mark Bishop. Abstract platforms of computation. In *Proceedings of the AISB Convention 2013*, pages 32–35, University of Exeter, 2013. Presented at the AISB Symposium on Computing and Philosophy: The Scandal of Computation - What is Computation? URL: https://centaur.reading.ac.uk/35696/1/Spencer%20et%20al.%20-%202013%20-%20Abstract%20platforms%20of%20computation.pdf.
332. Rupesh Kumar Srivastava, Klaus Greff, and Jürgen Schmidhuber. Training very deep networks. In *Advances in Neural Information Processing Systems 28 (NIPS 2015)*, pages 2377–2385, 2015. URL: https://proceedings.neurips.cc/paper/2015/file/215a71a12769b056c3c32e7299f1c5ed-Paper.pdf.
333. Erik Stayton and Jack Stilgoe. It's time to rethink levels of automation for self-driving vehicles. *IEEE Technology and Society Magazine*, 39(3):13–19, September 2020. URL: https://ieeexplore.ieee.org/document/9199318, doi:10.1109/MTS.2020.3012322.
334. Henrik Stewenius, Chris Engels, and David Nister. An efficient minimal solution for infinitesimal camera motion. In *2007 IEEE Conference on Computer Vision and Pattern Recognition*, pages 1–8, 2007. URL: https://ieeexplore.ieee.org/document/4270138, doi:10.1109/CVPR.2007.383113.
335. Robert Summers. Stochastic diffusion search: A basis for a model of visual attention. Master's thesis, Keele University, 1998. URL: https://www.doc.gold.ac.uk/~mas02mb/sdp/download/mscsds.pdf.
336. Ilya Sutskever, Oriol Vinyals, and Quoc V. Le. Sequence to sequence learning with neural networks, September 2014. 10th September 2014. URL: https://arxiv.org/abs/1409.3215, arXiv:arXiv:1409.3215.
337. Richard S. Sutton. The bitter lesson, March 2019. URL: https://www.cs.utexas.edu/~eunsol/courses/data/bitter_lesson.pdf.
338. Brad Templeton. Tesla, Waymo, Nuro, Zoox and many others embrace new AI to drive. *Forbes*, April 2024. 18th April 2024. URL: https://www.forbes.com/sites/bradtempleton/2024/04/18/tesla-waymo-nuro-zoox-and-many-others-embrace-new-ai-to-drive/.
339. Inc. Tesla. Q4 2024 earnings call transcript, 2025. URL: https://tinyurl.com/2jr2mxzv.
340. The Road to Autonomy. Tesla ushers in AV 2.0 era, March 2024. 31st March 2024. URL: https://www.roadtoautonomy.com/tesla-av/.

341. Evan Thompson. *Mind in Life: Biology, Phenomenology, and the Sciences of Mind*. Harvard University Press, Cambridge, MA, 2007.
342. Sebastian Thrun, Mike Montemerlo, Hendrik Dahlkamp, David Stavens, Andrei Aron, James Diebel, Philip Fong, John Gale, Morgan Halpenny, Gabriel Hoffmann, Kenny Lau, Celia Oakley, Mark Palatucci, Vaughan Pratt, and Pascal Stang. Stanley: The robot that won the DARPA grand challenge. *Journal of Field Robotics*, 23(9):661–692, September 2006. URL: https://robots.stanford.edu/papers/thrun.stanley05.pdf, doi:10.1002/rob.20147.
343. Kaitlyn Tiffany. George Hotz promised to end capitalism in a manic sermon at SXSW. *The Verge*, March 2016. 15th March 2016. URL: https://www.theverge.com/2016/3/15/11225314/george-hotz-self-driving-car-autopilot-tesla-speech-sxsw-2016.
344. Carlo Tomasi and Takeo Kanade. Shape and motion from image streams under orthography: A factorization method. *International Journal of Computer Vision*, 9:137–154, 1992. URL: https://link.springer.com/article/10.1007/BF00129684#citeas, doi:10.1007/BF00129684.
345. P. H. S. Torr and A. Zisserman. Feature based methods for structure and motion estimation. In Bill Triggs, Andrew Zisserman, and Richard Szeliski, editors, *Vision Algorithms: Theory and Practice*, pages 278–294, Berlin, Heidelberg, 2000. Springer Berlin Heidelberg. URL: https://link.springer.com/chapter/10.1007/3-540-44480-7_19.
346. Leonardo Torres y Quevedo. *My Inventions and Other Pages for General Dissemination*. Self-published, 1917.
347. Sadayuki Tsugawa, Teruo Yatabe, Takeshi Hirose, and Shuntetsu Matsumoto. An automobile with artificial intelligence. In *Proceedings of the International Joint Conference on Artificial Intelligence (IJCAI'79)*, IJCAI'79, pages 893–895, San Francisco, CA, USA, 1979. Morgan Kaufmann Publishers Inc. URL: https://dl.acm.org/doi/10.5555/1623050.1623117.
348. Alan Turing. Intelligent machinery (1948). In *The Essential Turing*. Oxford University Press, 09 2004. URL: https://academic.oup.com/book/42030/chapter-abstract/355746030?redirectedFrom=fulltext, doi:10.1093/oso/9780198250791.003.0016.
349. Alan M. Turing. On computable numbers, with an application to the Entscheidungsproblem. *Proceedings of the London Mathematical Society*, s2-42(1):230–265, 1937. URL: https://londmathsoc.onlinelibrary.wiley.com/doi/abs/10.1112/plms/s2-42.1.230, doi:10.1112/plms/s2-42.1.230.
350. Alan M. Turing. Computing machinery and intelligence. *Mind*, LIX(236):433–460, 1950. URL: https://academic.oup.com/mind/article/59/236/433/986238, doi:10.1093/mind/LIX.236.433.
351. Edwin L. Turner and Susan Schneider. Behavioral tests for AI consciousness, empathy, and goal content integrity, 2017.
352. B. Ulmer. Vita ii-active collision avoidance in real traffic. In *Proceedings of the Intelligent Vehicles Symposium IVS'94*, pages 1–6, 1994. URL: https://ieeexplore.ieee.org/document/639460/, doi:10.1109/IVS.1994.639460.

353. United Nations Economic Commission for Europe. UNECE paves the way for automated driving by updating UN international convention, 2016. URL: https://unece.org/press/unece-paves-way-automated-driving-updating-un-international-convention.
354. Urban Dictionary contributors. Cope, August 2024. URL: https://www.urbandictionary.com/define.php?term=cope.
355. V. N. Vapnik and A. Ya. Chervonenkis. On the uniform convergence of relative frequencies of events to their probabilities. *Theory of Probability & Its Applications*, 16(2):264–280, 1971. URL: https://epubs.siam.org/doi/10.1137/1116025, doi:10.1137/1116025.
356. Various. Machines can see 2023, May 2023. Accessed: 2025-01-10. URL: https://2023.machinescansee.com/.
357. Yanis Varoufakis. *Technofeudalism: What Killed Capitalism*. Penguin, 2023. URL: https://www.penguin.co.uk/books/451795/technofeudalism-by-varoufakis-yanis/9781529926095.
358. Ashish Vaswani, Noam Shazeer, Niki Parmar, Jakob Uszkoreit, Llion Jones, Aidan N. Gomez, Łukasz Kaiser, and Illia Polosukhin. Attention is all you need. In *Proceedings of the 31st International Conference on Neural Information Processing Systems*, NIPS'17, pages 6000–6010, Red Hook, NY, USA, 2017. Curran Associates Inc. URL: https://dl.acm.org/doi/10.5555/3295222.3295349.
359. Venkatkumar. Integrating diabetic retinopathy classification into webapp (gradio), February 2023. Accessed: 2025-01-08. URL: https://medium.com/@VK_Venkatkumar/integrating-diabetic-retinopathy-classification-into-webapp-gradio-977ef9ebbd4b.
360. Philip E. Vernon. *Intelligence and Attainment Tests*. University of London Press, 1960. URL: https://psycnet.apa.org/record/1961-03421-000.
361. Oriol Vinyals, Igor Babuschkin, Wojciech Marian Czarnecki, et al. Grandmaster level in starcraft ii using multi-agent reinforcement learning. *Nature*, 575(7782):350–354, 2019. URL: https://www.nature.com/articles/s41586-019-1724-z, doi:10.1038/s41586-019-1724-z.
362. Oriol Vinyals, Igor Babuschkin, Wojciech Marian Czarnecki, Michaël Mathieu, Andrew Dudzik, Junyoung Chung, David H. Choi, Richard Powell, Thomas Ewalds, Petko Georgiev, Junhyuk Oh, Dan Horgan, Marco Kroiss, Ivo Danihelka, John Agapiou, Max Jaderberg, David Silver, Demis Hassabis, Koray Kavukcuoglu, and Timothy Lillicrap. StarCraft ii: A new challenge for reinforcement learning, August 2017. URL: https://arxiv.org/abs/1708.04782, arXiv:arXiv:1708.04782.
363. John von Neumann. First draft of a report on the EDVAC. Technical report, Moore School of Electrical Engineering, University of Pennsylvania, June 1945. Contract No. W-670-ORD-4926 - Between the United States Army Ordnance Department and the University of Pennsylvania. URL: https://archive.org/details/20200901-vN_First_Draft_Report_EDVAC_Moore_Sch_1945/page/n1/mode/2up?view=theater.

364. A. Waibel, T. Hanazawa, G. Hinton, K. Shikano, and K.J. Lang. Phoneme recognition using time-delay neural networks. *IEEE Transactions on Acoustics, Speech, and Signal Processing*, 37(3):328–339, 1989. URL: https://ieeexplore.ieee.org/document/21701, doi:10.1109/29.21701.
365. Marsha Walton. Robots fail to complete grand challenge, May 2004. Accessed: 2025-01-03. URL: https://web.archive.org/web/20221129122602/http://www.cnn.com/2004/TECH/ptech/03/14/darpa.race/.
366. David Waltz. Understanding line drawings of scenes with shadows. In P. H. Winston, editor, *Psychology of Computer Vision*, pages 19–91. McGraw-Hill, New York, 1975.
367. Brian Wang. Why Waymo and Cruise will go bankrupt, August 2023. Accessed: 2025-01-08. URL: https://www.nextbigfuture.com/2023/08/why-waymo-and-cruise-will-go-bankrupt.html.
368. Pei Wang. What do you mean by 'AI'? In *Proceedings of the 2008 Conference on Artificial General Intelligence 2008: Proceedings of the First AGI Conference*, pages 362–373, NLD, 2008. IOS Press. URL: https://dl.acm.org/doi/10.5555/1566174.1566207.
369. Pei Wang. On defining artificial intelligence. *Journal of Artificial General Intelligence*, 10(2):1–37, 2019. URL: https://sciendo.com/article/10.2478/jagi-2019-0002, doi:10.2478/jagi-2019-0002.
370. Pei Wang. Intelligence: From definition to design. In Kristinn R. Thórisson, editor, *Proceedings of the Third International Workshop on Self-Supervised Learning*, volume 192 of *Proceedings of Machine Learning Research*, pages 35–47. PMLR, July 2022. URL: https://proceedings.mlr.press/v192/wang22a.html.
371. Qiang Wang, Bei Li, Tong Xiao, Jingbo Zhu, Changliang Li, Derek F. Wong, and Lidia S. Chao. Learning deep transformer models for machine translation, June 2019. URL: https://arxiv.org/abs/1906.01787, arXiv:arXiv:1906.01787.
372. Yixiao Wang, Chen Tang, Lingfeng Sun, Simone Rossi, Yichen Xie, Chensheng Peng, Thomas Hannagan, Stefano Sabatini, Nicola Poerio, Masayoshi Tomizuka, and Wei Zhan. Optimizing diffusion models for joint trajectory prediction and controllable generation, August 2024. URL: https://arxiv.org/abs/2408.00766, arXiv:arXiv:2408.00766.
373. Jason Wei, Xuezhi Wang, Dale Schuurmans, Maarten Bosma, Brian Ichter, Fei Xia, Ed H. Chi, Quoc V. Le, and Denny Zhou. Chain-of-thought prompting elicits reasoning in large language models, January 2022. URL: https://arxiv.org/abs/2201.11903#, arXiv:arXiv:2201.11903.
374. Blay Whitby. The Turing test: AI's biggest blind alley? In Peter Millican and Andy Clark, editors, *Machines and Thought: The Legacy of Alan Turing, Volume 1*, pages 53–62. Oxford University Press, 1996. URL: https://academic.oup.com/book/50139/chapter/422029419.
375. Alfred North Whitehead and Bertrand Russell. *Principia Mathematica*, volume 1. Cambridge University Press, Cambridge, UK, 1910. Subsequent volumes published in 1912 (Vol. 2) and 1913 (Vol. 3).

376. Ryan Whitwam. New study accuses lm arena of gaming its popular AI benchmark, May 2025. Accessed: 2025-05-13. URL: https://arstechnica.com/ai/2025/05/researchers-claim-lm-arenas-ai-leaderboard-is-biased-against-open-models/.
377. B. Widrow and M. E. Hoff. Adaptive switching circuits. In *IRE WESCON Convention Record*, pages 96–104, 1960.
378. Norbert Wiener. *Cybernetics, or Control and Communication in the Animal and the Machine.* Hermann & Cie, Paris, 1948. The original edition was simultaneously printed in the US by Wiley, offset from the Paris sheets.
379. Eugene P. Wigner. The unreasonable effectiveness of mathematics in the natural sciences. *Communications in Pure and Applied Mathematics*, 13(1):1–14, February 1960. URL: https://onlinelibrary.wiley.com/doi/abs/10.1002/cpa.3160130102.
380. Wikipedia. Jean piaget: Biological model of intellectual development. Wikipedia, 2025. Accessed: 2025-08-31. URL: https://en.wikipedia.org/wiki/Jean_Piaget.
381. Wikipedia contributors. Dead internet theory, 2024. URL: https://en.wikipedia.org/wiki/Dead_Internet_theory.
382. Wikipedia contributors. Ai slop, 2025. URL: https://en.wikipedia.org/wiki/AI_slop.
383. Terry Winograd. *A procedural model of language understanding*, pages 249–266. Morgan Kaufmann Publishers Inc., San Francisco, CA, USA, 1986. URL: https://dl.acm.org/doi/10.5555/21922.24337.
384. Ludwig Wittgenstein. *Tractatus Logico-Philosophicus.* Kegan Paul, Trench, Trubner & Co., London, 1922. Originally published in German as *Logisch-Philosophische Abhandlung* in 1921. URL: https://www.gutenberg.org/files/5740/5740-pdf.pdf.
385. Stephen Wolfram. *A New Kind of Science.* Wolfram Media, Inc., Champaign, IL, 2002. URL: https://www.wolframscience.com/nks/.
386. Stephen Wolfram. *What Is ChatGPT Doing ... and Why Does It Work?* Wolfram Media, Inc., February 2023. URL: https://writings.stephenwolfram.com/2023/02/what-is-chatgpt-doing-and-why-does-it-work/.
387. Han Wu, Syed Yunas, Sareh Rowlands, Wenjie Ruan, and Johan Wahlstrom. Adversarial driving: Attacking end-to-end autonomous driving. In *2023 IEEE Intelligent Vehicles Symposium (IV)*, pages 1–7, 2023. URL: https://ieeexplore.ieee.org/document/10186386.
388. Yoshio Yamaguchi, Hiroshi Kawai, Tetsuro Matsui, and Yoshihiro Kondo. A speaker-independent isolated word recognition system. In *Proceedings of the First International Conference on Spoken Language Processing (ICSLP '90)*, pages 1077–1080, November 1990. URL: https://www.isca-archive.org/icslp_1990/yamaguchi90c_icslp.html.
389. Brian Yang, Huangyuan Su, Nikolaos Gkanatsios, Tsung-Wei Ke, Ayush Jain, Jeff Schneider, and Katerina Fragkiadaki. Diffusion-ES: Gradient-free planning with diffusion for autonomous driving and zero-shot instruction

following, February 2024. URL: https://arxiv.org/abs/2402.06559, arXiv: arXiv:2402.06559.

390. Yinan Zheng, Ruiming Liang, Kexin Zheng, Jinliang Zheng, Liyuan Mao, Jianxiong L, Weihao Gu, Rui A, Shengbo Eben Li, Xianyuan Zhan, and Jingjing Liu. Diffusion-based planning for autonomous driving with flexible guidance. In *Proceedings of the International Conference on Learning Representations ICLR'25*, 2025. URL: https://openreview.net/pdf?id=wM2sfVgMDH.

391. Haoran Zhu, Zhenyuan Dong, Kristi Topollai, and Anna Choromanska. Ad-l-jepa: Self-supervised spatial world models with joint embedding predictive architecture for autonomous driving with lidar data, January 2025. URL: https://arxiv.org/abs/2501.04969, arXiv:arXiv:2501.04969.

392. John Zimmer. The third transportation revolution: Lyft's vision for the next ten years and beyond, September 2016. URL: https://medium.com/@johnzimmer/the-third-transportation-revolution-27860f05fa91.

Index

For Product Safety Concerns and Information please contact our EU representative GPSR@taylorandfrancis.com
Taylor & Francis Verlag GmbH, Kaufingerstraße 24, 80331 München, Germany

www.ingramcontent.com/pod-product-compliance
Ingram Content Group UK Ltd.
Pitfield, Milton Keynes, MK11 3LW, UK
UKHW022224190926
475702UK00003B/488

9781032911229